T0329604

Security and Privacy in the Internet of Things

Security and Privacy in the Internet of Things

Architectures, Techniques, and Applications

Edited by

Ali Ismail Awad
Luleå University of Technology, Luleå, Sweden

United Arab Emirates University, Al Ain, United Arab Emirates

Al-Azhar University, Qena, Egypt

Jemal Abawajy
Deakin University, Australia

IEEE PRESS
WILEY

Published by John Wiley & Sons, Inc., Hoboken, New Jersey.
Published simultaneously in Canada.

For general information on our other products and services or for technical support, please contact our Customer Care Department within the United States at (800) 762-2974, outside the United States at (317) 572-3993 or fax (317) 572-4002.
Wiley also publishes its books in a variety of electronic formats. Some content that appears in print may not be available in electronic formats. For more information about Wiley products, visit our web site at www.wiley.com.

Library of Congress Cataloging-in-Publication Data

Names: Awad, Ali Ismail, editor. | Abawajy, Jemal H., 1982- editor.
Title: Security and privacy in the Internet of things : architectures, techniques, and applications / edited by Ali Ismail Awad, Jemal Abawajy.
Other titles: Security and privacy in the Internet of things (Wiley-IEEE Press)
Description: Hoboken, NJ : Wiley-IEEE Press, [2022] | Includes bibliographical references and index.
Identifiers: LCCN 2021051556 (print) | LCCN 2021051557 (ebook) | ISBN 9781119607748 (hardback) | ISBN 9781119607762 (adobe pdf) | ISBN 9781119607779 (epub)
Subjects: LCSH: Internet of things–Security measures. | Data protection.
Classification: LCC TK5105.8857 .S444 2022 (print) | LCC TK5105.8857 (ebook) | DDC 005.8–dc23/eng/20211123
LC record available at https://lccn.loc.gov/2021051556
LC ebook record available at https://lccn.loc.gov/2021051557

Cover Design: Wiley
Cover Image: © Tuomas A. Lehtinen/Getty Images

Set in 9.5/12.5pt STIXTwoText by Straive, Chennai, India

10 9 8 7 6 5 4 3 2 1

Contents

About the Editors

 **Ali Ismail Awad, PhD**, is currently an Associate Professor with the College of Information Technology (CIT), United Arab Emirates University (UAEU), Al Ain, United Arab Emirates. He is also an Associate Professor (Docent) with the Department of Computer Science, Electrical and Space Engineering, Luleå University of Technology, Luleå, Sweden, where he also served as a coordinator of the Master Program in Information Security. He is an Associate Professor with the Electrical Engineering Department, Faculty of Engineering, Al-Azhar University at Qena, Qena, Egypt. He is also a Visiting Researcher with the University of Plymouth, United Kingdom. His research interests include Cybersecurity, the Internet of Things security, network security, image analysis with applications in biometrics, and medical imaging. He has edited or coedited seven books and authored or co-authored several journal articles and conference papers in these areas. He is an Editorial Board Member of the Future Generation Computer Systems Journal, Computers & Security Journal, the Internet of Things, Engineering Cyber Physical Human Systems Journal, Health Information Science and Systems Journal, and IET Image Processing Journal. Dr. Awad is currently an IEEE senior member.

Prof. Jemal Abawajy is on the executive team of the Centre for Cyber Security Research and Innovation (CSRI) leadership Team at Deakin University where he leads the Cyber Security and Technologies Division composed of three divisions (IoT/CPS and Critical Infrastructure Security, Digital Forensics and Incident Management, and Cyber Analytics and AI). Professor Abawajy's research is focused on both pure and applied research in the areas of distributed systems and network security (i.e. cloud security, fog security, Internet of Things security, blockchain, and big data security). He has published close to 400 refereed papers in these areas in top venues and his research is funded both by competitive grant and industry. He holds visiting Professorship in many universities. He has supervised over 40 PhD students and 20 Master's degree students to a successful completion. Currently he is supervising 10 PhD student and 1 Master student. He has also supervised over 15 postdoctoral candidates in the past five years. He is regularly invited to assess competitive grants for granting agencies in Europe (e.g. Swiss National Science Foundation), Asia (e.g. Agency for Science, Technology and Research), the Americas (e.g. Natural Sciences and Engineering Research Council of Canada), Australia (e.g. Australian Research Council), and Africa (e.g. South Africa's National Research Foundation). He is also regularly asked to assess degree programs for accreditations at many universities (e.g. Auckland University of Technology, Auckland, New Zealand, and University Putra Malaya), appointed on the Advisory Committee (e.g. Box Hill Institute and Imam Abdulrahman Bin Faisal University), and Academic Advisor (e.g. Universiti Kebangsaan Malaysia). Prof. Abawajy has served/serving on the editorial board of numerous international journals (e.g. IEEE Transaction on Cloud Computing and IEEE ACCESS) and more than 400 international academic conferences in various capacities including Chair (e.g. IEEE SmartCity-2020) and Program Chair (e.g. IEEE International Conference on Parallel and Distributed Systems). He has received numerous awards including Excellence in Research Award (e.g. Deakin University Vice-Chancellor's Award for Outstanding Contribution to Research), Best Paper Award (e.g. IEEE International Conference on Cloud Computing Technology and Science, Taiwan), and several service awards. Prof. Abawajy has given over 70 invited keynote speeches, and numerous invited seminars all over the world.

List of Contributors

Bamidele Adebisi
Department of Engineering
Manchester Metropolitan University
Manchester, UK

Adnan Anwar
School of Information Technology
Deakin University
Geelong
Australia

Zubair Baig
School of Information Technology
Deakin University
Geelong
Australia

Lejla Batina
Digital Security Group
iCIS, Radboud University
Nijmegen
The Netherlands

Aswani Kumar Cherukuri
School of Information Technology and
Engineering
Vellore Institute of Technology
Vellore
Tamil Nadu
India

Vinamra Das
School of Information Technology and
Engineering
Vellore Institute of Technology
Vellore
Tamil Nadu
India

Hugo Egerton
Department of Computing and
Mathematics
Manchester Metropolitan University
Manchester
UK

Steven Furnell
School of Computer Science
University of Nottingham
Nottingham
United Kingdom

and

Centre for Research in Information
and Cyber Security
Nelson Mandela University
Port Elizabeth
South Africa

Hassan Habibi Gharakheili
School of Electrical Engineering and
Telecommunications
University of New South Wales
Sydney
Australia

Saqib Hakak
Department of Computer Science
Faculty of Computer Science
University of New Brunswick
Information Technology Centre
NB E3B 5A3
Canada

Mohammad Hammoudeh
Department of Computing and
Mathematics
Manchester Metropolitan University
Manchester
UK

Ayyoob Hamza
School of Electrical Engineering and
Telecommunications
University of New South Wales
Sydney
Australia

Catherine Higgins
Department of Computer Science
Faculty of Computer Science
University of New Brunswick
Information Technology Centre
NB E3B 5A3
Canada

Muhammad Ijaz Ul Haq
Department of Computer Science
Faculty of Computer Science
University of New Brunswick
Information Technology Centre
NB E3B 5A3
Canada

Wenjuan Li
Department of Electronic and
Information Engineering
The Hong Kong Polytechnic University
Hong Kong
China

Gang Li
School of Information Technology
Deakin University
Geelong
Melbourne
Australia

Xiao Liu
School of Information Technology
Deakin University
Geelong
Melbourne
Australia

Martin Lundgren
Department of Computer Science
Electrical and Space Engineering
Luleå University of Technology
Luleå
Sweden

Lucas McDonald
Department of Computer Science
Faculty of Computer Science
University of New Brunswick
Information Technology Centre
NB E3B 5A3
Canada

Weizhi Meng
DTU Compute
Technical University of Denmark
Kgs. Lyngby
2800
Denmark

Nele Mentens
LIACS
Leiden University
Leiden
The Netherlands

and

imec-COSIC and ES&S
ESAT, KU Leuven
Leuven
Belgium

Markus Miettinen
System Security Lab
TU Darmstadt
Darmstadt
Germany

Naila Mukhtar
School of Engineering
Macquarie University
Sydney
Australia

Thien Duc Nguyen
System Security Lab
TU Darmstadt
Darmstadt
Germany

Ali Padyab
School of Informatics
University of Skövde
Skövde
Sweden

Md Masoom Rabbani
imec-COSIC and ES&S
ESAT, KU Leuven
Leuven
Belgium

Aditya Raj
School of Information Technology and
Engineering
Vellore Institute of Technology
Vellore
Tamil Nadu
India

Phillip Rieger
System Security Lab
TU Darmstadt
Darmstadt
Germany

Ahmad-Reza Sadeghi
System Security Lab
TU Darmstadt
Darmstadt
Germany

Abdulhadi Shoufan
Center for Cyber-Physical Systems
Khalifa University
Abu Dhabi
United Arab Emirates

Vijay Sivaraman
School of Electrical Engineering and
Telecommunications
University of New South Wales
Sydney
Australia

Sune Von Solms
Faculty of Engineering and the Built
Environment
University of Johannesburg
Johannesburg
South Africa

Naeem F. Syed
School of Science
Edith Cowan University
Perth
Australia

Bilal Taha
Department of Electrical and
Computer Engineering
University of Toronto
Toronto, ON
Canada

Ikram Sumaiya Thaseen
School of Information Technology and
Engineering
Vellore Institute of Technology
Vellore
Tamil Nadu
India

Devrim Unal
KINDI Center for Computing
Research
College of Engineering
Qatar University
Doha
Qatar

Chan Yeob Yeun
Center for Cyber-Physical Systems
Khalifa University
Abu Dhabi
United Arab Emirates

Preface

The Internet-of-Things (IoT) is an emerging paradigm due to extensive developments in information and communication technology (ICT). The purpose of IoT is to expand the functions of the first version of the Internet by increasing the ability to connect numerous objects. The IoT model has expanded to span different applications such as manufacturing and Industry 4.0, eHealth, smart cities and homes, robotics and drones, transportation, and critical infrastructures. The wide facilities offered by IoT and other sensing facilities have led to a huge amount of data generated from versatile domains; thus, security and privacy have become inevitable requirements not only for the sake of personal safety but also for assuring the sustainability of the IoT paradigm itself. Moreover, the nature and significance of the IoT systems themselves can increase their desirability as targets of attack. To get the full benefits of the IoT systems, the highest possible levels of security and privacy must be accomplished. However, as with the wide diversity of IoT applications and environments, several security and privacy issues remain unaddressed.

This book fills in the gaps in IoT security and privacy by providing the readers with cutting-edge research findings in the IoT security domain. This book outlines key emerging trends in IoT security and privacy considering the entire IoT architecture (perception, network, and applications) layers, with a focus on different critical IoT applications. The up-to-date body of knowledge presented in this book is a need for researchers, practitioners, and postgraduate students who work in the IoT development and deployment domains. This volume introduces a collection of 10 chapters written by experts in the field that cover both security and privacy aspects implied on IoT. Furthermore, the material has been prepared in a way that makes each chapter independently readable from the others, while still contributing a collective overall insight into the topic area. The book comprises 10 chapters structured as follows:

The book begins with the chapter *Advanced Attacks and Protection Mechanisms in IoT Devices and Networks,* authored by *Batina et al.*, which introduces a full picture of the possible attacks and the countermeasures spanning IoT perception and network layers. The chapter covers a wide spectrum of security attacks and countermeasures in the IoT paradigm and prepares the stage for a better understanding of security vulnerabilities and protection mechanisms. Physical attacks, profiling attacks, and IoT malware at the network level are covered and augmented by some real-world examples. Countermeasures like remote attestation, machine learning-based solutions, and the applications of deep learning and federated self-learning in anomaly detection are also covered. The chapter identifies some future research directions like employing Blockchain for solving IoT security challenges.

Humans form the weakest circle in the cybersecurity chain. Therefore, human-related and social security aspects should be taken into consideration in addition to the technical security solution. Chapter 2 titled *Human Aspects of IoT Security and Privacy*, written by *Solms and Furnell* considers the human aspects of security and privacy issues with particular focus upon the use of IoT in the domestic context, where the users are potentially the least prepared in terms of background knowledge and available support. The chapter examines the challenges that may be presented from the perspective of using and managing the range of IoT devices that are now to be found in smart home environments, and the related data storage and sharing that may be inherent in their use. The chapter demonstrates the need for user-facing security and privacy to receive comparable attention to that directed toward other elements of core functionality.

Back to the technical cybersecurity aspects, Chapter 3 named *Applying Zero Trust Security Principles to Defence Mechanisms Against Data Exfiltration Attacks*, authored by *Egerton et al.* describes data exfiltration threats that can emerge from within a company and external threats that seek to gain unauthorized access to sensitive information that could be used for personal gain or malicious purposes. Furthermore, the chapter presents a network-based mechanism that can mitigate the common physical attack methods that are used by malicious insiders. The chapter demonstrates that the network-based mechanism can defend against some network-level threats while also complementing existing security deployments.

Drones are becoming a key element in different applications and an integral part to facilitate people's lives. Yet, opening the airspace to drones will significantly increase the number of malicious users as well as cyber-physical attacks. Chapter 4 called *eSIM-based Authentication Protocol for UAV Remote Identification*, written by *Shoufan, Yeun, and Taha* proposes an authentic communication of drones' remote identification that is particularly relevant to the controlled operations of commercial and civil drones. The proposed authentication protocol ensures a secure communication of drone remote identification by employing an embedded Subscriber Identification Module (eSIM) where any user can verify the authenticity of a remote ID by using digital signatures. A security analysis of the proposed authentication protocol is presented along with formal verification using ProVerif.

In connection to security attacks and countermeasures, deep insights on the collaborative intrusion detection mechanism are given in Chapter 5 titled *Collaborative Intrusion Detection in the Era of IoT: Recent Advances and Challenges*, written by *Li and Meng*. The chapter starts by giving an overview of collaborative intrusion detection and then reports the cutting-edge research achievements in this direction. To build a complete picture, the authors highlight open challenges and limitations that can be considered for any future work related to the collaborative intrusion detection topic.

Network traffic is a rich source of information utilized by malware analysis and intrusion detection systems. IoT network traffic analysis has not received considerable attention yet. Chapter 6 authored by *Gharakheili, Hamza, and Siavaraman* and titled *Cyber-Securing IoT Infrastructure by Modeling Network Traffic* explores the privacy and security risks of IoT devices that can be systematically evaluated, demonstrating real-life threats to typical users posed by cyber attackers. Furthermore, the chapter presents a behavioral analysis of IoT network traffic that leads to the development of machine learning-based models for inferencing from flow-level network behavior of IoT devices. Lastly, flow-level inferencing models are trained for detecting anomalous patterns in network traffic of individual connected devices.

Encrypted network traffic analysis is an essential process to understand traffic characteristics and to identify normal and abnormal behaviors. Chapter 7 named *Integrity of IoT Network Flow Records in Encrypted Traffic Analytics*, written by *Cherukuri et al.* tackles the analysis of encrypted IoT traffic. The chapter proposes novel solutions for ensuring the integrity of the IoT traffic flow records. It proposes hashing and encryption-based mechanisms to address the integrity of flow records in encrypted traffic as a flow record authentication problem. Furthermore, the chapter demonstrates the proposed solutions in a simulated environment.

eHealth architectures are complex compositions of IoT devices. With the sophistication of a contemporary eHealth infrastructure, the benefits of technological advances can be reaped to render effective and efficient patient services. However, such a benefit is accompanied by an increasing cyber threat plan that has emerged owing to increasing connectivity between the eHealth devices, and the lack of proper mechanisms for ensuring the security of IoT devices of the eHealth system. Chapter 8 titled *Securing Contemporary eHealth Architectures: Techniques and Methods*, written by *Syed, Baig, and Anwar* presents a detailed overview of the security threats posed to eHealth systems, and the countermeasures thereof, with emphasis on IoT-enabled eHealth architectures.

Smart home devices have been adopted widely by household owners to fulfill a wide array of functions and needs. Investigating security and privacy issues from an end-user perspective, the deployment environment is often uncertain and basic security controls are lacking. Chapter 9 written by *Lundgren and Padyab* and titled *Security and Privacy of Smart Homes: Issues and Solutions*, proposes a user-centric model that furthers the research stream by extending the traditional CIA-triad from an enterprise-centric perspective to a user-centric privacy concerns perspective. The proposed model can help security professionals and developers to analyze smart home technology in terms of privacy violation consequences and concerns as perceived by end users.

Plenty of research endeavors to address several security challenges in IoT ecosystems have been conducted, but very little attention has been paid to the hardware-related security aspects of IoT devices. The final chapter of the book titled *IoT Hardware-Based Security: A Generalized Review of Threats and Countermeasures* written by *Hakak et al.* presents the status and concerns of hardware-based attacks on IoT devices. The chapter presents a general overview of IoT-based hardware attacks and discusses countermeasures that could be employed to mitigate or prevent those attacks and provides a forward-looking context of the IoT hardware security area.

The book, as a whole, documents the state-of-the-art, current challenges, cutting-edge research findings concerning IoT security and privacy areas with applications in different domains. The book is considered clear evidence of the research progress achieved in IoT security and privacy. Due to the rapid growth in the IoT paradigm, further contributions and findings are anticipated in this research domain. The presented contributions in this book will nonetheless highlight ideas and directions that can help in various circumstances, as well as supporting the holistic understanding of IoT security and privacy. As such, we hope that readers will find the book interesting and relevant as a contribution to the body of literature in this important area.

Ali Ismail Awad
Luleå University of Technology, Luleå, Sweden
United Arab Emirates University, Al Ain, UAE
Al-Azhar University, Qena, Egypt

Jemal Abawajy
Deakin University, Australia

1

Advanced Attacks and Protection Mechanisms in IoT Devices and Networks

Lejla Batina[1], Nele Mentens[2,3], Markus Miettinen[4], Naila Mukhtar[5], Thien Duc Nguyen[4], Md Masoom Rabbani[3], Phillip Rieger[4], and Ahmad-Reza Sadeghi[4]

[1]Digital Security Group, iCIS, Radboud University, Nijmegen, The Netherlands
[2]LIACS, Leiden University, Leiden, The Netherlands
[3]imec-COSIC and ES&S, ESAT, KU Leuven, Leuven, Belgium
[4]System Security Lab, TU Darmstadt, Darmstadt, Germany
[5]School of Engineering, Macquarie University, Sydney, Australia

Acronyms

CNN	Convolutional neural network
DDoS	Distributed denial of service
DoS	Denial of service
EM	Electromagnetic
FL	Federated learning
FN	False negative
FP	False positive
FPR	False positive rate
GRU	Gated recurrent unit
LSTM	Long short-term memory
ML	Machine learning
MLP	Multilayer perceptron
PDoS	Permanent denial of service
POI	Point of interest
ROC	Receiver operating characteristic
RNN	Recurrent neural network
SCA	Side-channel analysis

Security and Privacy in the Internet of Things: Architectures, Techniques, and Applications,
First Edition. Edited by Ali Ismail Awad and Jemal Abawajy.
© 2022 The Institute of Electrical and Electronics Engineers, Inc. Published 2022 by John Wiley & Sons, Inc.

SOHO	Small office/home office
TA	Template attack
TN	True negative
TP	True positive
TPR	True positive rate
WDDL	Wave dynamic differential logic
XOR	Exclusive OR

1.1 Introduction

In the past decade, the Internet of Things (IoT) has emerged as a wonder-pill to our problems. The low-cost, easy to use, and easy to deploy motto contributed to the explosion of IoT devices that permeate our surroundings. Today, the use cases of IoT systems are spread across different industries, notably, in oil and gas exploration, smart factories, smart homes, medical applications, military applications, etc. However, the tendency of "rushing" to the market often disregards proper testing and security features. These major flaws make IoT devices an easy prey to different cyberattacks. Since attackers exist with different capabilities, this chapter consists of three sections in which different threat models are assumed, as visualized in Figure 1.1:

- *Physical adversary.* Section 1.2 assumes that the attacker has physical access to the IoT device. It gives an overview of physical security threats and protection mechanisms of low-end IoT devices, focusing on side-channel analysis (SCA) and fault analysis attacks.
- *Remote software adversary.* Sections 1.3 and 1.4 assume that the attacker can remotely alter the software. Section 1.3 concentrates on remote attestation (RA) techniques, that aim at detecting malicious changes in the device's firmware by requesting a proof to verify the sanity of the device. Section 1.4 focuses on techniques to detect attacks by analyzing the data traffic in the IoT network.

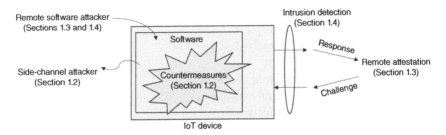

Figure 1.1 Attacks and protection mechanisms covered in this chapter.

1.2 Physical Security in IoT Devices

In this section, we first give an introduction to physical attacks, namely SCA and fault analysis attacks. Next, we discuss side-channel attacks that are based on profiling and on machine learning (ML) specifically. We continue the section by giving an overview of attacks on real-world devices. We conclude with discussing countermeasures against physical attacks.

The focus of this section is on low-end IoT devices. Microcontrollers and microprocessors including ARM, AVR, and MSP430, are commonly used platforms in this category. Identification of side-channel leakages from security algorithm implementations on any of these processor chips will potentially introduce vulnerabilities to all the IoT devices mounted with these chips.

1.2.1 Physical Attacks

One of the strongest attacker assumptions is that an adversary has physical access to an IoT device. The attacks that can be mounted in that case are called *physical attacks*. The first type of physical attacks, that we discuss in Section 1.2.1.1, are SCA attacks. The second type of physical attacks, i.e. fault analysis attacks, are discussed in Section 1.2.1.2. Note that this chapter does not cover side-channel attacks that exploit vulnerabilities in the micro-architecture of a processor. We concentrate on attacks based on weaknesses of the software or hardware implementation of cryptographic algorithms.

1.2.1.1 Side-channel Analysis Attacks

SCA attacks analyze the information of an electronic system available through side-channels, such as the power consumption, the electromagnetic (EM) emanation, or the timing behavior of the system, as shown on the right side of Figure 1.2. Whereas the main input/output channel should not leak secret information if the cryptographic algorithm is designed properly, the secret message or the secret key can be derived from side-channels if the algorithm is naively implemented. The first side-channel attack proposed in the academic community was a timing attack on public-key algorithms in 1996 by Paul Kocher [1]. The attack derives the secret exponent in a modular exponentiation, executed through repetitive square and multiply operations, by measuring the time it takes to perform (parts of) the exponentiation. A few years later, attacks that monitor the power consumption of an electronic system were proposed by Kocher et al. [2]. These attacks exploit the fact that the power consumption of a logic gate in a digital chip depends on the processed data. Most attacks are based on extracting information from the dynamic power consumption of the chip because that depends on the switching of the logic values internally in the chip. Indirectly measuring the same effect, attacks based on the EM emanation of electronic chips were introduced in 2001 by Gandolfi et al. [3] and Quisquater and Samyde [4].

1.2.1.2 Fault Analysis Attacks

Fault analysis attacks are based on deliberately inserted faults targeting the erroneous behavior of a hardware or software implementation. Boneh et al. were the first to show that cryptographic secrets can be leaked through the insertion of a fault [5]. On the one hand, researchers continued to find flaws in cryptographic algorithms in the form of leaked secret information in the presence of a fault. On the other hand, research is conducted into the physical insertion of faults. The examples given in this section and in Figure 1.2 are fault attacks based on laser injection, power glitches, and clock glitches. In 2002, Skorobogatov and Anderson showed that the value of a bit in a microcontroller can be changed using a camera flash and a laser pointer [6]. In the same year, Aumüller et al. demonstrated that underpowering a smart card for a short while can induce an exploitable fault [7]. Fukunaga and Takahashi show the effectiveness of introducing a glitch in the clock signal, i.e. shortening one period of the clock, to conduct a fault attack [8].

1.2.2 Profiling Attacks

Profiling attacks are regarded as one of the strongest side-channel attacks. In profiling attacks, the adversary has access to the cloned open copy of the device under target, also called the profiling device. Profiling attacks consist of two phases: the profiling (characterization or training) and attack (classification or matching) phase. In profiling phase, the attacker can program the profiling device to estimate a profile leakage model which is then utilized to recover the unknown secret key in the classification. Template attacks [9] and attacks based on stochastic models [10] are examples of profiling attacks.

Suppose an adversary has a cloned copy of the IoT device and is capable of capturing a number of leakage traces L_i, while the encryption E is performed on

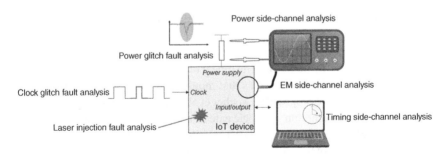

Figure 1.2 Side-channel and fault analysis attacks considered in this chapter.

the device with the key k^*. Let k^* (with $k^* \in K$) represent the fixed cryptographic key and t (with $t \in P$) represent the plaintext or ciphertext of the cryptographic algorithm. Then, y can be represented as the mapping of the plaintext or ciphertext t and key k^* to a value that is assumed to be related to the deterministic part of the measure leakage L, according to the model being used. Based on this, the measured leakage L can be represented (Eq. (1.1)) as a function of independent additive noise r and a device-specific function ϕ. This multivariate leakage $L = L_1, \ldots, L_N$ (where N represents the total number of leakage traces) is exploited in profiling attacks. So, in the profiling phase, the attacker has N leakage traces ($L = L_{P1}$, $L_{P2}, \ldots, L_{PN}$), collected by computing N encryptions using different plaintexts ($P = t_{P1}, t_{P2}, \ldots, t_{PN}$) and the same secret key $k *$. In the attack phase, Q traces ($L = L_{Q1}, L_{Q2}, \ldots, L_{QN}$) are collected by using different plaintexts and a different key. The traces in the attack phase are different from the traces in profiling phase.

$$L = \phi(y(t, k)) + r \qquad (1.1)$$

A template attack (TA) is theoretically the most powerful and commonly used side-channel attack. It relies on Bayes theorem and on the assumption that the leakage estimation function $L|(P, K)$ has a multivariate Gaussian distribution that is parameterized by its mean ($\mu_{p,k}$) and covariance ($\sum_{p,k}$), for each template (p, k). TA uses the generative model strategy which is used for classification in ML as well.

1.2.3 Machine Learning and SCA

ML-based systems have excelled in various domains by significantly improving the performance of applications. Over the past few years, the applicability of ML has been analyzed to efficiently recover secret information by exploiting side-channel leakage in various cryptographic algorithms including AES, ECC, RSA, 3DES, and lightweight cryptographic algorithms [11–14]. With the improved efficiency, ML-based SCA seems to surpass traditional SCA methods.

ML-SCA is an extension of profiling TAs in which the adversary first trains a model (profiling phase) and then launches the attack on the actual unknown test traces to recover the secret key (classification phase). ML can be used in supervised, unsupervised, or semi-supervised modes. In supervised, unsupervised, and semi-supervised learning modes, the model learns from datasets that consists of labeled leakage traces, unlabeled leakage traces, and a combination of leakage traces with and without labels, respectively. ML-based SCA concentrating on power consumption leakage was first introduced by Hospodar et al. [15], followed by a comparative study by Lerman et al. [16].

ML-based SCA can be further divided into a six-step methodology as explained in [17] and shown in Figure 1.3. Leakage traces L are collected from the target IoT

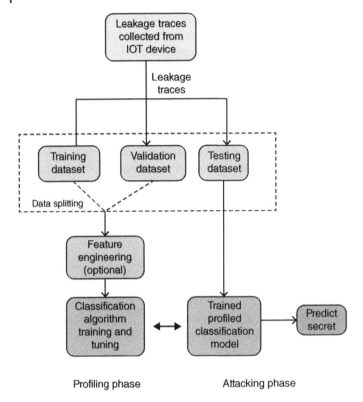

Figure 1.3 Machine learning-based evaluation methodology. Source: Adapted from Mukhtar et al. [17].

profiling device and are preprocessed to handle misalignment using filtering and windowing techniques. The traces are labeled according to the attack methodology. Each sample point in the leakage dataset represents a feature $(L_{i,f})$ and the label (C) represents the target key class. Each data trace is also referenced as a data instance in ML. The dataset consisting of aligned traces is then divided into three datasets: training, validation, and testing. The training dataset is a subset of the leakage traces dataset L, which are used to train the model. The validation and testing datasets consist of leakage traces which are used for validating the model during training and testing the model after training, respectively. The testing dataset is never shown to the model during training for an accurate efficient fitted model to avoid over-fitting. Over-fitting is a modeling error that occurs when the model closely fits on confined data instances, and fails to generalize on an unseen dataset. The redundant, insignificant features in the dataset might lower the model performance. Feature engineering techniques can be applied in case of simple ML analysis and are not required for deep learning analysis. The last step of the ML-based SCA methodology is the actual attack phase. The unseen traces are used to test the performance of the model. The best performing model is then used to predict the keys from unseen data traces.

Important factors for successful traditional side-channel attacks are alignment and significant feature or point of interest (POI) selection. In deep learning-based SCA, there is no need for preprocessing and alignment. Deep learning is a subset of ML and is popular thanks to its capability of self-learning from the data patterns using artificial neural networks without the requirement of preprocessing or feature engineering. This makes it practical from an attacker's perspective. In deep learning-based SCA, computational models are built which consist of multiple processing layers including an input layer, a series of hidden layers and an output layer [18]. The lucrative effortless secret key recovery analysis using deep learning neural networks opened up a new avenue of SCA research. Maghrebi et al. have presented results to recover secret information with neural networks [13]. Cagli et al. have demonstrated that ML-based SCA using convolutional neural networks (CNN) can help in neutralizing jitter-based hiding countermeasures without requiring any pre-processing, alignment or feature engineering [19]. Kim et al. proposed a method to add noise to the traces for a robust deep learning based SCA [20]. The visualization of the features and the hyperparameters can help in building an efficient neural network model [21].

IoT devices, processing sensitive information, have the potential risk of being exposed to physical threats. They can be physically captured, disassembled, and analyzed forensically to recover the secret information using side-channel leakage. The threats and damages are aggravated with the help of the ML paradigm. ML can be utilized to exploit side-channel leakages to investigate the internal software activities in IoT devices. This can introduce security vulnerabilities as well as can help in malicious activity or anomaly detection, proactively. Wang et al. presented results for the anomaly and software activity detection based on changes in EM leakages from IoT devices, and evaluated them using multilayer perceptrons (MLP), long short-term memory (LSTM) and autoencoders [22]. Sayakkara et al. presented a case of using ML-based SCA for forensic analysis to detect a wide variety of changes to the target IoT device to exploit secret information [23].

1.2.4 Real-world Attacks

Physical attacks exploiting weaknesses in implementations are a real threat for various kinds of IoT applications and devices including voice assistants, medical devices, self-driving cars, etc. Moreover, thanks to computing power becoming ever cheaper nowadays, modern adversaries have started using state-of-the-art ML algorithms for attacks.

The first side-channel attack in a real-world use case was performed on the implementations of the Keeloq algorithm, which was deployed in numerous car immobilizers and garage remote devices [24]. Other attacks followed, such as those on DESFire [25] and MIFARE classic implementations and even on ATMEL secure microcontrollers [26].

Considering attacks on largely deployed devices such as light bulbs, Ronen et al. showed how the key that Philips uses to encrypt and authenticate new firmware can be extracted using SCA, leading to a massive distributed denial of service (DDoS) attack [27]. Another real-world threat to a large class of IoT devices was demonstrated recently by a team of academics. In particular, their research showed that by shining a laser at microphones inside smart speakers, tablets, or phones, a far-away attacker can remotely send inaudible and potentially invisible commands into voice assistants, such as Google assistant, Amazon Alexa, and Apple Siri using light [28]. The attack was demonstrated at a distance of more than 50 m through the window. To prevent disasters, it is necessary to disable unauthorized access to all IoT devices and especially to medical ones. Gnad et al. performed a leakage assessment on three individual microcontrollers from two different vendors that are often used as IoT implementation platforms [29], and the work of Meulenaer and Standaert unveiled the threats of power analysis attacks to wireless sensor nodes [30].

1.2.5 Countermeasures

As mentioned above, the majority of real-world attacks on security implementations running on IoT devices today use SCA or fault injection to infer (secret) data or otherwise interfere with the devices' regular processing. Preventing this kind of attacks in general remains a great challenge, as effective mitigation measures are often prohibitively expensive in terms of power and energy resources.

Attacking implementations of security by using unintentionally leaked information has been revealed decades ago and ever since many countermeasures were proposed [31]. The most common approach for countermeasures aims at breaking the link between the information leaked through the side channel and the actual data that are being processed. One way to achieve it is to replace the real data by some other data (on which the computation is actually performed). To this principle, the literature usually refers to as masking or secret sharing [32] and it is the core idea behind threshold implementations (TI) where a bit is represented as a combination of several bits (e.g. using an exclusive OR operation or XOR) [33]. Another approach aims at breaking the link between the data manipulated by the device (in general) and the power consumed by the computations. This approach is called hiding, and one way to achieve it is by flattening the power consumption of a device by for example, using special logic styles that are more robust against

SCA attacks. An example is Wave Dynamic Differential Logic (WDDL) [34]. Note that this approach is specific for certain side channels, such as power consumption and EM emanations.

Fault injection, on the other hand should be countered by adding some form of redundancy, which will send a signal that a fault occurred before letting the adversary taking advantage of it. The approaches for this are often algorithm-specific resulting in recent research directions in designing fault resistant algorithms, moving the defenses consideration to the design phase [35].

1.3 Remote Attestation in IoT Devices

RA is a security protocol that runs between a trusted party called Verifier (**Vrf**) and "potentially" untrusted party called Prover (**Prv**).

A brief overview of the typical working procedure of RA is shown in Figure 1.4. During the bootstrapping phase, the **Vrf** and **Prv** are initialized with the key **K**. In Step 1, the **Vrf** sends a challenge C to the **Prv**, where C typically contains a Nonce to guarantee the freshness of the result. In Step 2, the **Prv** performs the attestation, i.e. a hashing operation of its underlying software, upon receiving the challenge. After that, the **Prv** computes the message authentication code (MAC) over the received C and the software digest and sends the response R to the **Vrf**. In Step 3, the **Vrf** computes R' using the stored golden value of the hash digest. In Step 4, upon receiving R, the **Vrf** compares R with R'. If both values match, the **Vrf** assumes that the **Prv** is in a "healthy" state, else the **Prv** is compromised.

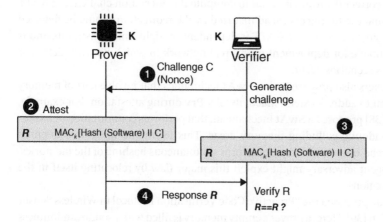

Figure 1.4 Typical example of classical remote attestation.

The type of adversary that RA intends to protect against, is a remote software adversary, as explained in Section 1.1. We distinguish two categories of remote software adversaries:

- *Static software adversary.* The main goal of a static software adversary is to introduce or run malicious code on a device at a specific location in the program memory. This type of adversary is described in [36].
- *Mobile software adversary.* This adversary is capable of infecting a device and changing its location within the IoT firmware to evade detection.

In what follows, we discuss different types of RA, from the device perspective (Section 1.3.1), and from the network perspective (Section 1.3.2). We conclude the section with an outlook to future directions (Section 1.3.3).

1.3.1 Types of Remote Attestation

Primarily, RA techniques can be subdivided into three main categories: (i) Software-based RA; (ii) Hardware-based RA, and (iii) Hybrid architecture-based RA. We described each of these techniques below.

1.3.1.1 Software-based Remote Attestation
As the name indicates, software-based Remote Attestation (**SwAt**) is based on software-based techniques and does not rely upon sophisticated hardware modules for the secure root of trust.

In general, **SwAt** schemes rely on a strict time-bound reply from the prover. As discussed in SWATT [37], a **Vrf** sends an attestation challenge to a **Prv**. The **Vrf** assumes that a compromised **Prv** can only reply with the correct attestation response if it redirects the attestation challenge to another proxy device that holds the correct firmware in order to compute the attestation challenge. SWATT assumes that the timing overhead incurred by the redirection should be detected during protocol execution. SWATT is dependent on tight time constraints and is thus unsuitable for deployment over a real network in which network delays or packet loss scenarios occur.

Researchers also proposed other **SwAt** techniques that involve part of memory verification to address extend inactivity of a **Prv** during attestation. For example, Spinellis [38] proposed a **SwAt** mechanism, that performs commutation of hashes of two randomly colluding memory areas. This technique involves sequential memory read-out and it does not perform simultaneous hashing of the memories. An intelligent adversary might exploit this major flaw by relocating itself in the memory regions.

Choi et al. proposed the "Proactive Code Verification Protocol in Wireless Sensor Network" in [39]. Here, a prover's empty memory is filled with pseudo randomness using a pseudo random function (PRF). During attestation, the **Vrf** sends a nonce to the **Prv** and the nonce acts as a seed for the PRF to generate pseudo random-ness. The main idea is to fill the empty memory region with pseudo randomness to prevent an adversary to evade detection by relocating itself. The **Prv** then computes the hash of the entire memory and sends the attestation result to the **Vrf**. The **Vrf** can also compute the hash and validate the received attestation result with the computed one.

Indeed, **SwAt** schemes are an interesting and low-cost solution. Thanks to their "hardware-less" approach, these schemes are easily employable. However, the unrealistic assumptions of strict time constraints and the lack of secure hardware support for cryptographic operations make them unsuitable for real network implementations.

1.3.1.2 Hardware-based Remote Attestation

Hardware-based Remote Attestation (**HwAt**) methods depend primarily on the usage of specialized hardware modules that act as a tamper-resistant secure root of trust. Over the past decades, researchers have proposed different techniques that rely upon sophisticated hardware modules. Below, we discuss a few of the **HwAt** schemes along with the pros and cons.

To verify the trustworthiness of a potentially untrusted **Prv**, Sailer et al. proposed to extend the functionality of the trusted platform module in [40]. The main goal of the proposed scheme is to maintain a sequence of trust which covers the application layer and the system configuration of the **Prv**. Before execution, all executables are measured and the measurement is protected by the trusted platform module (TPM). Moreover, the proposed **HwAt** can detect "unauthorized" code invocation.

Kil et al. proposed ReDAS (Remote Dynamic Attestation System) in [41]. The main essence of this idea is to extract the properties from an application source code. During code execution, all the activities (including unauthorized or malicious activities) are recorded and stored securely in the TPM of the **Prv**. At the time of attestation, the **Prv** sends the protected recorded values to the **Vrf**. Undoubtedly, this technique can identify malicious activities during code execution. However, ReDAS does not consider all the dynamic properties. Thus, an adversary can still evade detection.

More recently, Vliegen et al. propsed SACHa in [42]. SACHa is an field programmable gate array (FPGA) based self-attestation mechanism for embedded platforms. As FPGAs are reconfigurable after deployment, this feature makes them inherently susceptible to attacks. Thus, it is indeed crucial to perform attestation of not only the software but also the hardware of an FPGA based system. By performing partial reconfigurability of an FPGA, SACHa performs a self-attestation of the FPGA system. By using SACHa, an FPGA can be used as a trusted platform module in the absence of a dedicated trusted hardware platform.

In summary, a **HwAt** scheme provides better security in terms of protecting the cryptographic details from adversaries and unauthorized code access. However, a trusted platform module is a costly hardware tool and often disregarded in IoT systems due to the cost and the IoT platform's inability to accommodate the platform.

1.3.1.3 Hybrid Architecture-based Remote Attestation

To cope with the demand of the IoT realm, researchers proposed software/hardware-based hybrid architectures that are low-cost and suitable for IoT deployment. In this type of system, the hybrid architecture acts as a secure root of trust for attestation.

El Defrawy et al. propose SMART in [43], a software/hardware co-design for low-end embedded devices that guarantees a dynamic root of trust for RA. The main idea of this architecture is to facilitate a secure memory location for attestation related details, for example, attestation code and an attestation key. The modified processor, to which minimal changes in the form of these secure memory locations have been applied, protects the secure memory locations from unauthorized "non-SMART" codes. That guarantees uninterrupted and secure attestation operation.

Other hybrid architecture based attestation schemes, such as TrustLite [44] and TyTan [45], also proposed similar approaches. Both TrustLite and Tytan are based on an Execution Aware Memory Protection Unit (EA-MPU). An EA-MPU enforces code-specific memory access and prevents unauthorized code access along with secure inter-process communication.

1.3.2 Remote Attestation for Large IoT Networks

So far, we discussed attestation schemes that are based on one-to-one device settings, where a **Vrf** can perform attestation over a single **Prv** at a time. However, these techniques do not scale. Hence, they are unsuitable for a large network. To address these issues, researchers have proposed "swarm-attestation" techniques, that can perform attestation over a considerably large IoT network in an acceptable time limit.

1.3.2.1 Classical Swarm Attestation Techniques

Swarm attestation techniques aim to address the scalability challenge of a classical RA technique. Typically, a swarm attestation scheme relies upon a hybrid architecture for a secure root of trust. Broadly speaking, swarm attestation techniques are classified into two main categories; (i) swarm attestation in static networks and (ii) swarm attestation in a dynamic networks.

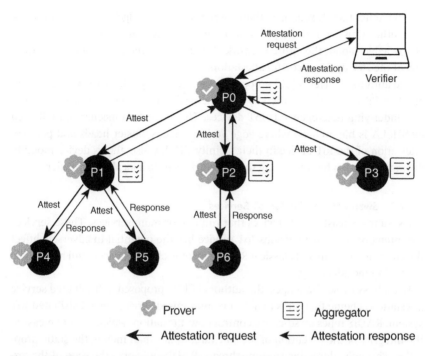

Figure 1.5 Typical example of classical swarm attestation scheme.

Swarm Attestation in Static Networks Swarm attestation schemes such as SEDA [46], SANA [47], and LISA [48] assume that the entire network is interconnected and static during the entire attestation duration. The network forms an overlay of a spanning tree rooted at the **Vrf**. The attestation process is distributed and parent nodes perform the attestation of the children nodes, as shown in Figure 1.5. Upon receiving the attestation results of their children, the parent nodes aggregate their own result and send it to their respective parent nodes. This mechanism is efficient in terms of runtime; however, the strong assumption of network connectivity for the entire attestation period makes it unsuitable for dynamic and intermittent networks.

Swarm Attestation in Dynamic Networks To facilitate the attestation of dynamic swarms, the authors of PADS [49] proposed a novel attestation technique based on a consensus mechanism. The essence of PADS is that every node performs a "self-attestation" (i.e. every node is assumed to be equipped with a hybrid architecture that contains a root of trust and minimal secure hardware to perform the attestation) and stores the result in the secure memory module. As the provers are dynamic, they "exchange" their respective "knowledge" about the network with other provers. Provers perform minimum-consensus to agree on common "knowledge" about the entire network. A **Vrf** can get the attestation result from any of the **Prv** upon successful attestation.

The authors of SHeLA [50] proposed a beyond state-of-the-art RA scheme that employs FPGAs as edge verifiers by introducing them between the root verifier and the underlying heterogeneous IoT devices. The attestation mechanism followed by SHeLA is hierarchical where edge verifiers act as cluster heads and perform attestation of the IoT device in their vicinity. SHeLA also allows device mobility during attestation by connecting the mobile devices to a local edge verifier.

1.3.2.2 Swarm Attestation for IoT Services

IoT systems consist of IoT devices that run one or more services. These services are interconnected. A malicious IoT service has the potential to adversely affect the entire IoT system and classical RA, which only checks devices, will not be able to identify the adversary.

To address these challenges, the authors of [51] proposed a distributed service attestation scheme that aims to detect compromised services in a distributed IoT system. RADIS relies on so-called control flow attestation, which aims to measure the integrity of the execution of the software, rather than merely the static properties. Through addressing runtime threats, RADIS enhances the state-of-the-art. Nevertheless, the need to precompute all possible control flows, not only within one device but also for interactions between devices in the network, together with the need for secure trusted hardware modules, makes RADIS difficult to scale and mostly suitable for small and controlled IoT environments such as smart homes.

Different from RADIS, the authors of SARA [52] proposed an RA scheme that performs attestation over a network that adopts the publish/subscribe communication pattern. SARA stores the historical evidence about service communication along with the code execution. In this way, a **Vrf** in SARA can construct all the historical communications that lead to an event. Note that SARA not only assumes synchronous communication but also asynchronous communication among IoT services. Based on the constructed communication chain, a **Vrf** can identify remote software adversaries in the network.

1.3.3 Future Directions

Indeed, RA techniques solve many issues; however, the current state-of-the-art falls short in many areas. Several new paradigms should be explored in the field of RA. We provide some major areas below.

1.3.3.1   Cloud-based RA Techniques

In the past decade, cloud/fog computing has emerged as the backbone of different industries by providing potentially "infinite" support in terms of software, hardware, storage, application, data analysis, etc. However, in RA, the exploitation of the cloud/fog infrastructure is often overlooked. A RA technique is an energy-draining process for many low-end embedded devices. Nevertheless, RA could be achieved by exploiting cloud computations on behalf of those "tiny" devices.

1.3.3.2   RA in Novel Internet Technologies

Recently, different new Internet architectures have been proposed, e.g. ICN [53], NDN [54], and SCIONLAB [55]. Researchers proposed many IoT applications that exploits these architectures. Unfortunately, none of the RA schemes exploits these architectures. Future RA research should exploit the built-in features such as in-network caching to address open issues like intermittent connectivity during attestation.

1.3.3.3   Blockchain Based RA

Traditionally, RA is a centralized process, where a central authority performs the attestation of IoT devices. However, this centralized system may not be suitable for many distributed IoT systems. To deal with this issue, blockchain technology can be applied. This will not only provide distributed secure storage but also a publicly verifiable distributed root of trust. The use of blockchain technology in RA may also solve the issues of public verifiability and allow any device to join or leave the network at runtime.

1.4   Intrusion Detection in IoT Networks

In this section, we will discuss challenges that IoT malware poses on IoT systems and introduce state-of-the-art intrusion detection approaches for IoT networks for coping with such threats.

1.4.1   IoT Malware

Recently, the number of deployed IoT devices and manufacturers providing IoT devices has been increasing rapidly. Fierce competition among manufacturers emphasizes the need for individual players to bring their devices very quickly to the market and save in development costs so that relatively little time and effort can be devoted to considering security aspects of the design and implementation of devices thoroughly. Consequently, many devices are released containing inherent security vulnerabilities that make them vulnerable to attacks. This has made many IoT devices a tempting target for attacks and has led to the emergence of so-called IoT malware, a new class of malicious software specifically targeting IoT devices. IoT malware has been responsible for numerous large-scale attacks, as reported by, e.g. [56–58]. For example the Mirai malware [56] exploits basic security flaws such as open Telnet access with default easily-guessed credentials to compromise devices. It has infected hundreds of thousands of IoT devices and used them to launch in 2016 the largest DDoS attack known by that time [56, 59]. Since then a number of new IoT malware variants have emerged including *Persirai* [60], *Hajime* [57, 61], *BrickerBot* [58], and *Silex* [62]. The analysis of [63] and [64] has shown that IoT malware attacks typically consist of four stages: *intrusion, infection, scanning*, and *exploitation or monetization*. In the intrusion stage, the malware obtains access to the victim device by utilizing vulnerabilities or weaknesses like default credentials. In the infection stage, the malware uploads the malware binary using a built-in file transfer protocol, e.g. `tftp` to the device and executes it. After infection, the device typically scans the network to find potential other vulnerable devices. In the exploitation stage (also referred to as monetization), the malware performs malicious actions as instructed by the adversary like executing DDoS attacks against particular victim destinations. For example, the Mirai malware supports ten different denial of service (DoS) attack vectors [65]: `udp` (UDP flood), `syn` (SYN flood), `ack` (ACK flood), `stomp` (TCP stomp flood), `udpplain` (UDP flood with less options), `vse` (Valve source engine specific flood), `dns` (DNS resolver flood), `greip` (GRE IP flood), `greeth` (GRE Ethernet flood), and `http` (HTTP flood). Moreover, malware like BrickerBot [58] causes a *permanent denial-of-service* (PDoS) by deleting crucial data on the compromised device, thus rendering the device permanently unusable. For this particular case, detecting attacks at an early stage is crucially important.

In Section 1.4.2, we will introduce state-of-the-art defenses against IoT malware and related open research issues.

1.4.2 Vulnerability Patching

Current best-practice protection methods against IoT malware are often not sufficient. The most common and preferable approach would be to roll out software update patches for found vulnerabilities (see e.g. [66]) and thereby eliminating them. Two main issues make this method ineffective in the IoT context: (i) many IoT devices might not be equipped with facilities for automatic updating and (ii) there is a significant delay from the moment a vulnerability is discovered and exploitable until the manufacturer releases a patch addressing the issue thus leaving the device vulnerable for extended periods of time. An alternative solution could be using an intrusion detection system (IDS) to detect and stop attacks. However, this approach also has several drawbacks as discussed below.

1.4.3 Signature- and Anomaly Detection-based Network Intrusion Detection

Signature-based intrusion detection is a well-established technique in network intrusion detection, in which the system detects attacks by comparing observed traffic to known attack patterns [67]. Its drawback is, however, that it cannot detect new or unknown attacks, i.e. it requires IDS providers to identify novel attacks and update their attack signature databases with corresponding signatures. A recent study from [63] shows that around 83% of IoT malware samples (88 samples out of a total 106 of samples) captured by their honeypot were not covered by the Virus Total database.[1] Moreover, there is a great dynamism in the recent growth of IoT malware threat vectors. According to a study by the anti-virus vendor Kaspersky, there were about 120,000 IoT malware samples detected in the first half of 2018 alone, triple the number for the whole year 2017. Due to this steep growth in threats, it would be inefficient to solve the problem through the use of predefined attack signatures. An alternative method for detecting attacks that does not suffer from this problem is anomaly detection that does not require pre-defined patterns for performing detection (see e.g. [68–70]). This method relies on learning a profile representing normal device behaviors and detecting any abnormal behavior deviating from these profiles. The problem with anomaly detection is, however, that in order to be effective, the method often generates many false alarms in which normal behavior is classified as abnormal. This is because it is difficult to capture the set of all normal behaviors accurately in a single profile. Due to this, anomaly detection-based IDS systems presented by Aqil et al. [71] and Nobakht et al. [72] only achieve a false positive rate (*FPR*) of 9.1 % or 5.8 %, respectively, which is by far too much for any practical system.

A better approach is therefore required in order to be able to address the ever-increasing number of device categories. In addition, making the problem even more challenging, many IoT devices usually do not generate a lot of network traffic, resulting in a lack of data for training detection models effectively. To address these challenges, we will present a new promising type of IDS introduced by Nguyen et al. [73] that leverages device-type specific deep neural network models to learn the normal network traffic patterns of IoT devices and to detect attack patterns that deviate from the device type's behavior.

1.4.4 Deep Learning-based Anomaly Detection

Several anomaly detection-based IDSs have been proposed. CIoTA, a host-based anomaly detection system proposed by Golomb et al. [74] analyses the control flow of applications running on IoT devices by using a Markov Chain. In the approach

---

1 https://www.virustotal.com

of Rahman et al. in [75], multiple clients train collaboratively a neural network to classify the properties of traffic, like the applied network protocols or flags. They use a technique called federated learning (FL) to merge training results from different clients. We will discuss FL later in detail. Portnoy et al. [69] cluster all network packets, also using the properties of the packets and consider the majority clusters as benign traffic. In this book, however, we provide a detailed discussion of DÏoT [73], one of the most prominent approaches that has drawn a lot of attention (see e.g. [76–79]). DÏoT is based on a neural network using GRUs (gated recurrent unit) for anomaly detection. It considers network packet sequences in a similar way as strings in a natural language and uses this abstraction to profile the packet sequence patterns that each IoT device of a specific type uses in their communication. The whole of modeling and evaluating communication patterns is completely automatic and the approach achieves high accuracy compared to other state-of-the-art approaches by e.g. Aqil et al. [71] or Nobakht et al. [72]. More interestingly, it is the first approach that leverages *FL* to enable clients to jointly train a global detection model without actually have to share their data, i.e. individual clients' data are always kept locally. Thus, this provides privacy and efficiency benefits as we will discuss in Section 1.4.5. Next, we will explain this approach in more detail.

1.4.4.1 System Overview

An anomaly detection-based IDS for IoT has three types of entities (see e.g. [73, 75]):

- *IoT devices.* The devices that the system aims to protect by monitoring and analyzing their traffic data.
- *SGW (security gateway).* The SGWs act as the local WiFi routers in end-user networks, so that all IoT devices, e.g. in the smart home of a user connect to the Internet through the SGW over WiFi or an Ethernet connection. In this way, the gateway is able to monitor all communications of IoT devices in its network and detect anomalous communications.
- *IoTSS (IoT security service).* In centralized learning, IoTSS collects data from SGWs and trains the detection model. In FL, its main role is to coordinate SGWs to jointly train the detection model.

1.4.4.2 Modeling Packet Sequence Patterns

The core idea of profiling traffic patterns of DÏoT is to transform a packet sequence $\langle pk_1, pk_2, \ldots, pk_n \rangle$ of benign traffic generated by an IoT device into sequences of symbols $\langle s_1, s_2, \ldots, s_n \rangle$, in which each packet pk_i is mapped to a symbol s_i based on its distinct characteristics. To learn benign traffic patterns, DÏoT employs GRUs (Chung et al. [80]), an emerging recurrent neural network (RNN), to train the detection model using the symbol sequences to represent the packet sequence patterns. Moreover, DÏoT assigns each distinct device type a dedicated GRU model to learn the normal packet sequences of packets present in the communication patterns of each device type. To detect abnormal traffic patterns, DÏoT feeds the symbols representing packet flows to the trained GRU model and estimates the occurrence probability of each new packet given the sequence of preceding packets. If the occurrence probability estimates p_i of a sufficient number of consecutive packets fall below a detection threshold, this packet sequence is considered anomalous and an alarm is raised as discussed in detail in Section 1.4.4.3.

In order to map network traffic packets pk_i into symbols $s_i = (c_1, c_2, \ldots, c_7)$, DÏoT uses seven packet characteristics as seven-tuples of discrete values:

- c_1 *direction (incoming/outgoing traffic)*. Normal TCP communication is usually two-way balanced, but abnormal is not, e.g. in DoS attacks, where a bot constantly sends packets to a victim without receiving responses.
- c_2 *(local) and* c_3 *(remote) port type*. These are the bin indices of three port types: system, user, and dynamic. In normal traffic, the ports depend on a specific device's implementation and the network protocol used. In attack traffic, however, ports are associated with the specific network protocols that the adversary aims to attack. For example, the adversary must use ports 80 or 443 for http-based DDoS attacks.
- c_4 *packet length*. Denotes the bin index of the eight most frequent packet length values with an additional bin index for all other packet length values.
- c_5 *TCP flags*. The TCP flag values range from 1 to 255. Normal TCP communication follows certain flag sequences like SYN, ACK, and then PUSH, etc. Attack traffic often does not follow such patterns, e.g. in a SYN flood, a type of DoS attacks, bots only sends TCP packets with the SYN flag.
- c_6 *protocols*. These are the encapsulated network protocol types. The set of specific protocols of a device type usually depends on the functionality of the device and associated design choices of the manufacturer. These protocols are usually different from the protocols used in attacks.
- c_7 *inter-arrival time (IAT) bin*. Denotes is the bin index of packet inter-arrival time. DÏoT uses three bins: $\langle$ 0.001, 0.001, to 0.05, and $\rangle$ 0.05 ms to distinguish normal traffic from high-bandwidth attack traffic used by, e.g. DoS attacks.

1.4.4.3 Anomalous Packet Detection
The anomaly detection process is performed by the SGWs using the trained GRU models for different device types provided by the traffic modeling and training steps explained above. The network packet flow is transformed into its corresponding symbol sequence as discussed above and then sequentially fed to the detection model. For each symbol s_i of the sequence, the detection model will output

an *occurrence probability* p_i based on the given sequence of k preceding symbols $\langle s_{i-k}, s_{i-k+1}, \ldots, s_{i-1} \rangle$, i.e. $p_i = P(s_i | \langle s_{i-k}, s_{i-k+1}, \ldots, s_{i-1} \rangle)$, where parameter k indicates the number of preceding symbols, i.e. the length of the lookback history that the GRU uses when calculating probability estimates for the possible next symbols in the symbol sequence. As a result, the GRU model will provide for each packet in the observed sequence of incoming packets an estimate of how likely the occurrence of this packet is given the preceding sequence of k packets. DÏoT shows that these probability estimates for known benign symbol sequence patterns will be on average higher than the ones for anomalous symbol sequence patterns, possibly generated by malware. In particular, it defines a packet to be *anomalous*, if the occurrence probability of the symbol corresponding to it is below a predefined *detection threshold T*, i.e. $p_i < T$. However, since benign traffic also contains noise that may not be captured by the GRU, triggering an alarm each time an anomalous packet is identified would lead to numerous false positives. Therefore, DÏoT aggregates the detection results over several consecutive packets and triggers an alarm only in the case that a significant number of packets within a sliding window are anomalous.

1.4.5 Federated Deep Learning-based IoT Intrusion Detection System

1.4.5.1 Federated Learning

FL is a distributed ML approach, in which clients collaborate to train a joint model using their own data [73, 81]. An FL system consists of K clients $(C_1, \ldots, C_K)$ and an aggregator $\mathcal{S}$. Let W_k denote a local model trained by client C_k and G the global model aggregated from local models $W_1, \ldots, W_K$. In a training round t, each client k receives the global model from previous training round G_{t-1} and uses its own data to retrain the model in order to obtain a new updated local model W_k. All local models $(W_1, \ldots, W_K)$ are sent to the aggregator to be aggregated into a new global model G_t using an aggregation algorithm. The most commonly used algorithm is FedAvg proposed by McMahan et al. [81], in which the global model is calculated by averaging the local models and weighting each model by the number of data samples used to train it. Let n_k denote the number of data samples of client k and n be the total number of samples of all clients, i.e. $n = \sum_{k=1}^{K} n_k$. We formalize the calculation of G_t in Eq. (1.2):

$$G = \sum_{k=1}^{K} \frac{n_k}{n} W_k \tag{1.2}$$

1.4.5.2 Federated Self-Learning Anomaly Detection

To train deep learning anomaly detection models, a sufficient number of observations of the normal behavior of IoT devices is required. This is a challenge since IoT devices typically are simple single-use appliances like motion sensors or smart coffee machines and thus usually generate only little data communications. Therefore, to obtain a sufficient amount of data for training comprehensive deep neural network-based anomaly detection models without delay, it is necessary to *aggregate* communications collected from several IoT networks. However, exporting these data to a central entity to be used for training poses a threat, since IoT device communications can be used to infer privacy-sensitive information, about particular user actions in the local networks [82, 83]. In order to tackle this challenge, several FL-based IoT IDSs have been proposed, e.g. by Nguyen et al. [73] and Rahman et al. [75]. Typically, these approaches enable clients (SGWs) to train the detection models (called local models) locally using their own data collected from the IoT devices in their networks. The local models are sent to the IoTSS to be aggregated into a global model using an aggregation algorithm, e.g. FedAvg as mentioned above. In this book, we take the DÏoT system introduced above as an example of such system. As presented in Section 1.4.4, DÏoT uses neural network-based models to detect compromised IoT devices in local networks. In this section, we will explain how the system leverages a FL-based training approach. The overall system setup is shown in Figure 1.6. It consists of a number of local SGWs, which collaborate with IoTSS to train anomaly detection models based on GRUs (a type of RNN), which were introduced by Chung et al. [80], for detecting anomalous behavior of IoT devices. To train detection models used for anomaly detection in a federated way, SGWs locally train detection models, then send them to the IoTSS, who aggregates them to a global model and propagates this global model back to the SGWs. Therefore, each SGW can

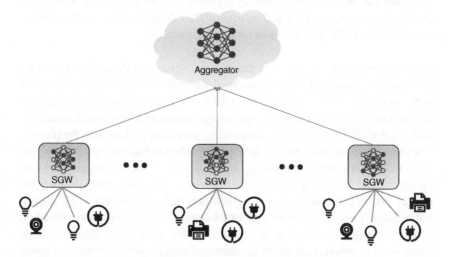

Figure 1.6 Overview of the FL-based IoT intrusion detection system.

benefit from training data contributed by all participating gateways. During the operation of the system, the training of the detection model is repeated in order to gradually increase the accuracy of the model, as more training data become available. This repeated training process is performed either routinely or until the global model reaches a specific level of convergence, i.e. when the model doesn't improve significantly anymore.

1.4.5.3 Challenges of Federated Learning-based Anomaly Detection System for IoT

Poisoning attacks. FL is an attractive target for so-called poisoning attacks, since the training process is distributed and local training is not controlled by the aggregator in any way. These attacks can be roughly divided into two classes: *untargeted poisoning* or DoS attacks, e.g. proposed by Fang et al. [84], Blanchard et al. [85], and Mu noz-González et al. [86], as well as *targeted poisoning* (or backdoor attacks), as proposed by Shen et al. [87], Bagdasaryan et al. [88], and Fung et al. [89]. Untargeted poisoning attacks aim to make the model unusable by sabotaging its overall classification performance, while backdoor attacks aim to inject a specific "backdoor" into the global model providing adversary-chosen incorrect outputs for specific inputs. To do that the adversary manipulates the local training process, e.g. uses poisoned data or crafts the local model's parameters directly and sends the poisoned model to the aggregator where it is aggregated into the global model. As the result, the malicious functionalities of the local poisoned model of the adversary are also incorporated in the global model. Nguyen et al. [90] show that FL-based IDSs for IoT are also vulnerable to poisoning attacks, in which the adversary can poison training data by gradually injecting malicious traffic such that the system cannot detect it. Consequently, the detection model will be trained to learn attack traffic patterns as normal traffic patterns.

Inference Attacks In an inference attack, an adversary analyzes the models for learning information about the data used for training them (see e.g. [91] and [92]). In the FL setting, since the aggregator can access all local model updates (cf. Section 1.4.5.1), an *honest-but-curious aggregator* can perform such attacks on each local model to learn the corresponding participant's data used for training that model without accessing the data directly. The fundamental goal of applying FL in IoT IDS is to protect the privacy of the data of IoT users by keeping the data locally instead of sending them to the aggregator. However, through inference attacks [91, 92], the aggregator can still learn information about the data of individual IoT users. To defend against this attack, several approaches based on secure multi-computation (e.g. [93]) or differential privacy (e.g. [94]) have been introduced. However, state-of-the-art inference attacks show that those defenses are insufficient (see e.g. [95]).

References

1 Kocher, P.C. (1996). Timing attacks on implementations of Diffie-Hellman, RSA, DSS, and other systems. In: *Advances in Cryptology — CRYPTO '96* (ed. N. Koblitz), 104–113. Berlin Heidelberg: Springer-Verlag.

2 Kocher, P., Jaffe, J., and Jun, B. (1999). Differential power analysis. *Annual International Cryptology Conference*, pp. 388–397. Springer.

3 Gandolfi, K., Mourtel, C., and Olivier, F. (2001). Electromagnetic analysis: concrete results. *International Workshop on Cryptographic Hardware and Embedded Systems*, pp. 251–261. Springer.

4 Quisquater, J.-J. and Samyde, D. (2001). Electromagnetic analysis (EMA): measures and counter-measures for smart cards. *International Conference on Research in Smart Cards*, pp. 200–210. Springer.

5 Boneh, D., DeMillo, R.A., and Lipton, R.J. (1997). On the importance of checking cryptographic protocols for faults. *International Conference on the Theory and Applications of Cryptographic Techniques*, pp. 37–51. Springer.

6 Skorobogatov, S.P. and Anderson, R.J. (2002). Optical fault induction attacks. *International Workshop on Cryptographic Hardware and Embedded Systems*, pp. 2–12. Springer.

7 Aumüller, C., Bier, P., Fischer, W. et al. (2002). Fault attacks on RSA with CRT: concrete results and practical countermeasures. *International Workshop on Cryptographic Hardware and Embedded Systems*, pp. 260–275. Springer.

8 Fukunaga, T. and Takahashi, J. (2009). Practical fault attack on a cryptographic LSI with ISO/IEC 18033-3 block ciphers. *2009 Workshop on Fault Diagnosis and Tolerance in Cryptography (FDTC)*, pp. 84–92. IEEE.

9 Chari, S., Rao, J.R., and Rohatgi, P. (2003). Template attacks. In: *Cryptographic Hardware and Embedded Systems - CHES 2002* (ed. B.S. Kaliski, ç.K. Kocc, and C. Paar), 13–28. Berlin, Heidelberg: Springer-Verlag.

10 Schindler, W., Lemke, K., and Paar, C. (2005). A stochastic model for differential side channel cryptanalysis. In: *Cryptographic Hardware and Embedded Systems – CHES 2005* (ed. J.R. Rao and B. Sunar), 30–46. Berlin, Heidelberg: Springer-Verlag. ISBN 978-3-540-31940-5.

11 Markowitch, O., Lerman, L., and Bontempi, G. (2011). Side channel attack: an approach based on machine learning. *Constructive Side-Channel Analysis and Secure Design, COSADE*.

12 Gilmore, R., Hanley, N., and O'Neill, M. (2015). Neural network based attack on a masked implementation of AES. *2015 IEEE International Symposium on Hardware Oriented Security and Trust (HOST)*, 106–111. Institute of Electrical and Electronics Engineers (IEEE), 5 2015. https://doi.org/10.1109/HST.2015.7140247.

13 Maghrebi, H., Portigliatti, T., and Prouff, E. (2016). Breaking cryptographic implementations using deep learning techniques, pp. 3–26, 12 2016. ISBN 978-3-319-49444-9. https://doi.org/10.1007/978-3-319-49445-6_1.

14 Carbone, M., Conin, V., Cornélie, M.-A. et al. (2019). Deep learning to evaluate secure RSA implementations. *IACR Transactions on Cryptographic Hardware and Embedded Systems* 2019 (2): 132–161. https://doi.org/10.13154/tches. v2019.i2.132-161.

15 Hospodar, G., Gierlichs, B., De Mulder, E. et al. (2011). Machine learning in side-channel analysis: a first study. *Journal of Cryptographic Engineering* 1 (4): 293. https://doi.org/10.1007/s13389-011-0023-x.

16 Lerman, L., Poussier, R., Bontempi, G. et al. (2015). Template attacks vs. machine learning revisited and the curse of dimensionality in side-channel analysis. In: *Revised Selected Papers of the 6th International Workshop on Constructive Side-Channel Analysis and Secure Design - Volume 9064, COSADE 2015*, pp. 20–33. Berlin, Heidelberg: Springer-Verlag. https://doi.org/10.1007/ 978-3-319-21476-4_2.

17 Mukhtar, N., Papachristodoulou, L., Fournaris, A.P. et al. (2020). Machine-learning assisted side-channel attacks on RNS-based elliptic curve implementations using hybrid feature engineering. https://eprint.iacr.org/2020/1065 (accessed 16 September 2021).

18 LeCun, Y., Haffner, P., Bottou, L., and Bengio, Y. (1999). Object recognition with gradient-based learning. In: *Shape, Contour and Grouping in Computer Vision*, 319. Berlin, Heidelberg: Springer-Verlag. ISBN 3540667229.

19 Cagli, E., Dumas, C., and Prouff, E. (2017). Convolutional neural networks with data augmentation against jitter-based countermeasures. *Cryptographic Hardware and Embedded Systems - CHES 2017 - 19th International Conference*, pp. 45–68, 08 2017. ISBN 978-3-319-66786-7. https://doi.org/10.1007/ 978-3-319-66787-4_3.

20 Kim, J., Picek, S., Heuser, A. et al. (2019). Make some noise. Unleashing the power of convolutional neural networks for profiled side-channel analysis. *IACR Transactions on Cryptographic Hardware and Embedded Systems* 2019 (3): 148–179, https://doi.org/10.13154/tches.v2019.i3.148-179.

21 Zaid, G., Bossuet, L., Habrard, A., and Venelli, A. (2019). Methodology for efficient CNN architectures in profiling attacks. *IACR Transactions on Cryptographic Hardware and Embedded Systems* 2020 (1): 1–36. https://doi.org/10.13154/tches.v2020.i1.1-36.

22 Wang, X., Zhou, Q., Harer, J. et al. (2018). Deep learning-based classification and anomaly detection of side-channel signals. In: *Cyber Sensing 2018*, vol. 10630 (ed. I.V. Ternovskiy and P. Chin), 37–44. International Society for Optics and Photonics, SPIE. https://doi.org/10.1117/12.2311329.

23 Sayakkara, A., Le-Khac, N.-A., and Scanlon, M. (2019). Leveraging electromagnetic side-channel analysis for the investigation of IoT devices. *Digital Investigation* 29: S94–S103. https://doi.org/10.1016/j.diin.2019.04.012.

24 Eisenbarth, T., Kasper, T., Moradi, A. et al. (2008). On the power of power analysis in the real world: a complete break of the KeeLoq code hopping scheme. In: *CRYPTO, Lecture Notes in Computer Science*, (ed. David Wagner), vol. 5157, 203–220. Santa Barbara, CA: Springer.

25 Oswald, D. and Paar, C. (2011). Breaking Mifare DESFire MF3ICD40: power analysis and templates in the real world. In *CHES*, pp. 207–222.

26 Balasch, J., Gierlichs, B., Verdult, R. et al. (2012). Power analysis of atmel CryptoMemory - RECOVERING KEYS FROM SECURE EEPROMs. In: *Topics in Cryptology - CT-RSA 2012 - The Cryptographers' Track at the RSA Conference 2012, San Francisco, CA, USA, February 27 - March 2, 2012. Proceedings, Lecture Notes in Computer Science*, vol. 7178 (ed. O. Dunkelman, 19–34. Springer. https://doi.org/10.1007/978-3-642-27954-6_2.

27 Ronen, E., Shamir, A., Weingarten, A.-O., and O'Flynn, C. (2018). IoT goes nuclear: creating a zigbee chain reaction. *IEEE Security and Privacy* 16 (1): 54–62. https://doi.org/10.1109/MSP.2018.1331033.

28 Sugawara, T., Cyr, B., Rampazzi, S. et al. (2020). Light commands: laser-based audio injection attacks on voice-controllable systems. In: *29th USENIX Security Symposium, USENIX Security 2020, August 12–14, 2020* (ed. S. Capkun and F. Roesner), 2631–2648. USENIX Association. https://www.usenix.org/conference/usenixsecurity20/presentation/sugawara.

29 Gnad, D.R.E., Krautter, J., and Tahoori, M.B. (2019). Leaky noise: new side-channel attack vectors in mixed-signal IoT devices. *IACR Transactions on Cryptographic Hardware and Embedded Systems* (3): 305–339. https://doi.org/10.13154/tches.v2019.i3.305-339.

30 de Meulenaer, G. and Standaert, F.-X. (2010). Stealthy compromise of wireless sensor nodes with power analysis attacks. In: *Mobile Lightweight Wireless Systems* (ed. P. Chatzimisios, C. Verikoukis, I. Santamaria et al.), 229–242. Berlin, Heidelberg: Springer-Verlag. ISBN 978-3-642-16644-0.

31 Mangard, S., Oswald, E., and Popp, T. (2006). *Power Analysis Attacks: Revealing the Secrets of Smart Cards*. Springer. http://www.springer.com/, ISBN 0-387-30857-1, http://www.dpabook.org/.

32 Chari, S., Jutla, C.S., Rao, J.R., and Rohatgi, P. (1999). Towards sound approaches to counteract power-analysis attacks. In: *CRYPTO*, (ed. Michael Wiener), *LNCS*, vol. 1666. Santa Barbara, CA: Springer. ISBN: 3-540-66347-9.

33 Nikova, S., Rijmen, V., and Schläffer, M. (2011). Secure hardware implementation of nonlinear functions in the presence of glitches. *Journal of Cryptology* 24 (2): 292–321.

34 Tiri, K., Hwang, D., Hodjat, A. et al. (2005). Prototype IC with WDDL and differential routing – DPA resistance assessment. In: *Proceedings of CHES'05*, (ed. Josyula R. Rao and Berk Sunar), *LNCS*, vol. 3659, 354–365. Edinburgh, Scotland: Springer.

35 Simon, T., Batina, L., Daemen, J. et al. (2020). Friet: An authenticated encryption scheme with built-in fault detection. In: *Advances in Cryptology -*

EUROCRYPT 2020 - *39th Annual International Conference on the Theory and Applications of Cryptographic Techniques, Zagreb, Croatia, May 10-14, 2020, Proceedings, Part I, Lecture Notes in Computer Science*, vol. 12105 (ed. A. Canteaut and Y. Ishai), 581–611. Springer. https://doi.org/10.1007/978-3-030-45721-1_21.

36 Abera, T., Asokan, N., Davi, L. et al. (2016). Invited – things, trouble, trust: on building trust in IoT systems. *Proceedings of the 53rd Annual Design Automation Conference*, p. 121. ACM.

37 Seshadri, A., Perrig, A., Van Doorn, L., and Khosla, P. (2004). SWATT: Software-based attestation for embedded devices. *Proceedings of the 2004 IEEE Symposium on Security & Privacy*, IEEE S&P '04, pp. 272–282.

38 Spinellis, D. (2000). Reflection as a mechanism for software integrity verification. *ACM Transactions on Information and System Security* 3 (1): 51–62.

39 Choi, Y.-G., Kang, J., and Nyang, D. (2007). Proactive code verification protocol in Wireless Sensor Network. *ICCSA'07*, pp. 1085–1096. Springer-Verlag.

40 Sailer, R., Zhang, X., Jaeger, T., and van Doorn, L. (2004). Design and implementation of a TCG-based integrity measurement architecture. *Proceedings of the 13th Conference on USENIX Security Symposium*, Volume 13, SSYM'04, p. 16.

41 Kil, C., Sezer, E.C., Azab, A.M. et al. (2009). Remote attestation to dynamic system properties: towards providing complete system integrity evidence. *2009 IEEE/IFIP DSN'09*, pp. 115–124.

42 Vliegen, J., Rabbani, M.M., Conti, M., and Mentens, N. (2019). SACHa: Self-attestation of configurable hardware. *2019 Design, Automation Test in Europe Conference Exhibition (DATE)*, pp. 746–751.

43 El Defrawy, K., Tsudik, G., Francillon, A., and Perito, D. (2012). SMART: Secure and minimal architecture for (establishing dynamic) root of trust. *Proceedings of the 19th Annual Network & Distributed System Security Symposium*, NDSS '12, pp. 1–15.

44 Koeberl, P., Schulz, S., Sadeghi, A.-R., and Varadharajan, V. (2014). TrustLite: A security architecture for tiny embedded devices. *Proceedings of the 9th European Conference on Computer Systems*, EuroSys '14, p. 10.

45 Brasser, F., El Mahjoub, B., Sadeghi, A.-R. et al. (2015). TyTAN: Tiny trust anchor for tiny devices. *Proceedings of the 52nd Design Automation Conference*, DAC '15, pp. 1–6.

46 Asokan, N., Brasser, F., Ibrahim, A. et al. (2015). SEDA: Scalable embedded device attestation. *Proceedings of the 22nd ACM SIGSAC Conference on Computer and Communications Security*, CCS '15, pp. 964–975.

47 Ambrosin, M., Conti, M., Ibrahim, A. et al. (2016). SANA: Secure and scalable aggregate network attestation. *Proceedings of the 2016 ACM SIGSAC Conference on Computer and Communications Security*, CCS '16, pp. 731–742.

48 Carpent, X., El Defrawy, K., Rattanavipanon, N., and Tsudik, G. (2017). Lightweight swarm attestation: a tale of two LISA-s. *Proceedings of the 2017 ACM on Asia Conference on Computer and Communications Security*, ASIACCS '17, pp. 86–100. ACM.

49 Ambrosin, M., Conti, M., Lazzeretti, R. et al. (2018). PADS: Practical attestation for highly dynamic swarm topologies. *Proceedings of the 1st International Workshop on Secure Internet of Things*, InSIoT '18.

50 Rabbani, M.M., Vliegen, J., Winderickx, J. et al. (2019). SHeLA: Scalable heterogeneous layered attestation. *IEEE Internet of Things Journal* 6 (6): 10240–10250. https://doi.org/10.1109/JIOT.2019.2936988.

51 Conti, M., Dushku, E., and Mancini, L.V. (2019). RADIS: Remote attestation of distributed IoT services. *2019 6th International Conference on Software Defined Systems*, SDS '19, pp. 25–32. IEEE.

52 Dushku, E., Rabbani, M.M., Conti, M. et al. (2020). SARA: Secure asynchronous remote attestation for IoT systems. *IEEE Transactions on Information Forensics and Security* 15: 3123–3136.

53 Fang, C., Yao, H., Wang, Z. et al. (2018). A survey of mobile information-centric networking: research issues and challenges. *IEEE Communication Surveys and Tutorials* 20 (3): 2353–2371. https://doi.org/10.1109/COMST.2018.2809670.

54 Zhang, L., Afanasyev, A., Burke, J. et al. (2014). Named data networking. *ACM SIGCOMM Computer Communication Review* 44 (3): 66–73.

55 Kwon, J., García-Pardo, J.A., Legner, M. et al. (2020). SCIONLab: A next-generation internet testbed. *Proceedings of the 28th IEEE International Conference on Network Protocols (ICNP)*. /publications/papers/icnp2020_scionlab.pdf.

56 Antonakakis, M., April, T., Bailey, M. et al. (2017). Understanding the Mirai Botnet. *26th USENIX Security Symposium (USENIX Security 17)*, pp. 1093–1110, Vancouver, BC. USENIX Association. ISBN 978-1-931971-40-9.

57 Herwig, S., Harvey, K., Hughey, G. et al. (2019). Measurement and analysis of Hajime, a peer-to-peer IoT Botnet. *26th Annual Network and Distributed System Security Symposium, NDSS*, 2019, San Diego, California, USA (24–27 February 2019). https://www.ndss-symposium.org/ndss-paper/measurement-and-analysis-of-hajime-a-peer-to-peer-iot-botnet/ (accessed 17 September 2021).

58 Radware (2017). "BrickerBot" results in PDoS attack. https://security.radware .com/ddos-threats-attacks/brickerbot-pdos-permanent-denial-of-service/ (accessed 17 September 2021).

59 Krebs, B. (2016). KrebsOnSecurity hit with record DDoS. https:// krebsonsecurity.com/2016/09/krebsonsecurity-hit-with-record-ddos/ (accessed 17 September 2021).

60 Yeh, T., Chiu, D., and Lu, K. (2017). Persirai: New internet of things (IoT) botnet targets IP cameras. https://blog.trendmicro.com/trendlabs-security-intelligence/persirai-new-internet-things-iot-botnet-targets-ip-cameras/ (accessed 17 September 2021).

61 Edwards, S. and Profetis, I. (2016). Hajime: Analysis of a Decentralized Internet worm for IoT Devices. Technical report No. 16. Rapidity Networks.

62 ZDNet (2019). New Silex malware is bricking IoT devices, has scary plans. https://www.zdnet.com/article/new-silex-malware-is-bricking-iot-devices-has-scary-plans/ (accessed 17 September 2021).

63 Minn, Y., Pa, P., Suzuki, S. et al. (2016). IoTPOT: A novel honeypot for revealing current IoT threats. *Journal of Information Processing* 24 (3): 522–533.

64 Kolias, C., Kambourakis, G., Stavrou, A., and Voas, J. (2017). DDoS in the IoT: Mirai and other Botnets. *Computer* 50 (7): 80–84.

65 Gamblin, J. (2020). Mirai source code. https://github.com/jgamblin/Mirai-Source-Code (accessed 17 September 2021).

66 Hadar, N., Siboni, S., and Elovici, Y. (2017). A lightweight vulnerability mitigation framework for IoT devices. *Proceedings of the 2017 Workshop on Internet of Things Security and Privacy*, IoTS&P '17, pp. 71–75, New York, USA: ACM. ISBN 978-1-4503-5396-0, https://doi.org/10.1145/3139937.3139944.

67 Doshi, R., Apthorpe, N., and Feamster, N. (2018). Machine learning DDoS detection for consumer Internet of Things devices. *CoRR*, abs/1804.04159. http://arxiv.org/abs/1804.04159.

68 Krügel, C., Toth, T., and Kirda, E. (2002). Service specific anomaly detection for network intrusion detection. *Proceedings of the 2002 ACM Symposium on Applied Computing*, pp. 201–208. ACM.

69 Portnoy, L., Eskin, E., and Stolfo, S. (2001). Intrusion detection with unlabeled data using clustering. *ACM CSS Workshop on Data Mining Applied to Security (DMSA-2001*. CiteSeer.

70 Rajasegarar, S., Leckie, C., and Palaniswami, M. (2014). Hyperspherical cluster based distributed anomaly detection in wireless sensor networks. *Journal of Parallel and Distributed Computing* 74 (1): 1833–1847.

71 Aqil, A., Khalil, K., Atya, A.O.F. et al. (2017). Jaal: Towards network intrusion detection at ISP scale. *Proceedings of the 13th International Conference on Emerging Networking EXperiments and Technologies*, CoNEXT '17, pp. 134–146. New York, USA: ACM. ISBN 978-1-4503-5422-6. https://doi.org/10.1145/3143361.3143399.

72 Nobakht, M., Sivaraman, V., and Boreli, R. (2016). A host-based intrusion detection and mitigation framework for smart home IoT using openflow. *Proceedings of 11th International Conference on Availability, Reliability and Security*, ARES 2016. IEEE.

73 Nguyen, T.D., Marchal, S., Miettinen, M. et al. (2019). DÏoT: A federated self-learning anomaly detection system for IoT. *The 39th IEEE International Conference on Distributed Computing Systems (ICDCS).*

74 Golomb, T., Mirsky, Y., and Elovici, Y. (2018). CIoTA: Collaborative IoT anomaly detection via blockchain. *Decentralized IoT Systems and Security, A Workshop in Conjunction with NDSS.*

75 Rahman, S.A., Tout, H., Talhi, C., and Mourad, A. (2020). Internet of things intrusion detection: centralized, on-device, or federated learning? *IEEE Network.*

76 Neshenko, N., Bou-Harb, E., Crichigno, J. et al. (2019). Demystifying IoT security: an exhaustive survey on IoT vulnerabilities and a first empirical look on internet-scale IoT exploitations. *IEEE Communication Surveys and Tutorials* 21 (3): 2702–2733.

77 Marchal, S., Miettinen, M., Nguyen, T.D. et al. (2019). AuDI: Towards autonomous IoT device-type identification using periodic communication. *IEEE Journal on Selected Areas in Communications* 1. https://doi.org/10.1109/JSAC.2019.2904364.

78 Afek, Y., Bremler-Barr, A., Hay, D. et al. (2020). NFV-based IoT security for home networks using MUD. *NOMS 2020-2020 IEEE/IFIP Network Operations and Management Symposium*, pp. 1–9. IEEE.

79 Koroniotis, N., Moustafa, N., and Sitnikova, E. (2019). Forensics and deep learning mechanisms for botnets in internet of things: a survey of challenges and solutions. *IEEE Access* 7: 61764–61785.

80 Chung, J., Gülçehre, c.C., Cho, K.H., and Bengio, Y. (2014). Empirical evaluation of gated recurrent neural networks on sequence modeling. *CoRR*, abs/1412.3555. http://arxiv.org/abs/1412.3555.

81 McMahan, B., Moore, E., Ramage, D. et al. (2017). Communication-efficient learning of deep networks from decentralized data. *Proceedings of the 20th International Conference on Artificial Intelligence and Statistics, AISTATS 2017*, Fort Lauderdale, FL, USA (20–22 April 2017), pp. 1273–1282. http://proceedings.mlr.press/v54/mcmahan17a.html (accessed 17 September 2021).

82 Acar, A., Fereidooni, H., Abera, T. et al. (2020). Peek-a-Boo: I see your smart home activities, even encrypted! *WiSec 2020: 13th ACM Conference on Security and Privacy in Wireless and Mobile Networks*, pp. 207–218. http://tubiblio.ulb.tu-darmstadt.de/121669/ (accessed 17 September 2021).

83 Apthorpe, N., Reisman, D., and Feamster, N. (2017). A smart home is no castle: privacy vulnerabilities of encrypted IoT traffic. *arXiv preprint arXiv:1705.06805.*

84 Fang, M., Cao, X., Jia, J., and Gong, N.Z. (2019). Local model poisoning attacks to Byzantine-Robust federated learning. *To appear in Usenix Security Symposium 2020.*

85 Blanchard, P., El Mhamdi, E.M., Guerraoui, R., and Stainer, J. (2017). Machine learning with adversaries: Byzantine tolerant gradient descent. In: *Advances in Neural Information Processing Systems, NIPS,* 119–129. Curran Associates, Inc.

86 Mu noz-González, L., Co, K.T., and Lupu, E.C. (2019). Byzantine-Robust Federated Machine Learning through Adaptive Model Averaging. *arXiv e-prints,* art. arXiv:1909.05125.

87 Shen, S., Tople, S., and Saxena, P. (2016). Auror: Defending against poisoning attacks in collaborative deep learning systems. *Proceedings of the 32nd Annual Conference on Computer Security Applications,* ACSAC '16, pp. 508–519. ACM.

88 Bagdasaryan, E., Veit, A., Hua, Y. et al. (2020). How to backdoor federated learning. *AISTATS.* PMLR.

89 Fung, C., Yoon, C.J.M., and Beschastnikh, I. (2018). Mitigating Sybils in federated learning poisoning. *CoRR,* abs/1808.04866. http://arxiv.org/abs/1808 .04866.

90 Nguyen, T.D., Rieger, P., Miettinen, M., and Sadeghi, A.-R. (2020). Poisoning attacks on federated learning-based IoT intrusion detection system. *Decentralized IoT Systems and Security, A Workshop in Conjunction with NDSS 2020.*

91 Pyrgelis, A., Troncoso, C., and Cristofaro, E.D. (2018). Knock knock, who's there? Membership inference on aggregate location data. *NDSS.*

92 Ganju, K., Wang, Q., Yang, W. et al. (2018). Property inference attacks on fully connected neural networks using permutation invariant representations. *CCS.*

93 Bonawitz, K., Ivanov, V., Kreuter, B. et al. (2017). Practical secure aggregation for privacy-preserving machine learning. *Proceedings of the 2017 ACM SIGSAC Conference on Computer and Communications Security,* CCS '17, pp. 1175–1191. New York, USA: ACM. ISBN 978-1-4503-4946-8. https://doi.org/10.1145/3133956.3133982.

94 McMahan, H.B., Ramage, D., Talwar, K., and Zhang, L. (2018). Learning differentially private language models without losing accuracy. *6th International Conference on Learning Representations.* http://arxiv.org/abs/1710.06963 (accessed 17 September 2021).

95 Melis, L., Song, C., De Cristofaro, E., and Shmatikov, V. (2019). Exploiting unintended feature leakage in collaborative learning. *2019 IEEE Symposium on Security and Privacy (SP),* pp. 691–706. https://doi.org/10.1109/SP.2019.00029.

2

Human Aspects of IoT Security and Privacy

Sune Von Solms[1] and Steven Furnell[2,3]

[1]*Faculty of Engineering and the Built Environment, University of Johannesburg, Johannesburg, South Africa*
[2]*School of Computer Science, University of Nottingham, Nottingham, United Kingdom*
[3]*Centre for Research in Information and Cyber Security, Nelson Mandela University, Port Elizabeth, South Africa*

2.1 Introduction

There has been a significant shift in the domestic digital environment, with smart home technologies enabling interactions to become more "ambient" and users less strictly tied to personal hardware. However, this shift has not yet been met with an appropriate shift toward recognition of the security and privacy issues that they can usher in for the users concerned. Individuals typically share access to various common devices such as smart TVs, speakers, appliances, smart sensors, and associated hub devices. However, these users are typically unaware of how their individual data is shared and stored in the IoT ecosystem and have little opportunity to understand and manage their environment from this perspective. Their perception of the devices as *appliances* rather than as computers exacerbates the challenge of ensuring that security and protection are given appropriate consideration.

In the wider context, the adoption and integration of IoT technologies span several areas and is seen to bringing opportunities in a range of contexts, including [1]

- Smart manufacturing where factory and logistics solutions are used to optimize processes, controls, and quality.
- Smart cities where IoT innovations aim to improve the quality of life in cities, addressing issues such as security and energy resourcefulness.
- Smart mobility which encompasses real-time route management and solutions aimed at making travel more enjoyable and transportation more reliable.

Security and Privacy in the Internet of Things: Architectures, Techniques, and Applications,
First Edition. Edited by Ali Ismail Awad and Jemal Abawajy.
© 2022 The Institute of Electrical and Electronics Engineers, Inc. Published 2022 by John Wiley & Sons, Inc.

- Smart life which includes innovative, state-of-the-art IoT technology with the aim to make life simpler and safer for the individual.

The mention of factors such as quality of life, enjoyment, and safety all implicitly highlights that while the IoT is ultimately based upon *technology* devices, it typically exists to serve a very *human* purpose. However, this is not necessarily apparent in the way the topic is presented and discussed. For example, the IEEE started a project in 2015 to create a comprehensive definition of IoT to facilitate a better understanding of the subject. They concluded, however, that a comprehensive definition of Internet of Things (IoT) is not easy as the used definition often focuses on the entities and assets deemed relevant for a specific application [2]. Research conducted on various IoT definitions and descriptions supports this finding, where the majority of IoT definitions largely focus on the technology and the potential that it brings. Definitions focus first on the technical configuration of complex network which interconnects various "things" to the Internet. Second, it focuses on the sensing and actuation capabilities of the devices, and the services which can be offered, with or without human intervention, based on the intelligent processing on the data captured by the devices. Few definitions mention security and privacy, and where it is mentioned it is listed simply as an added phrase at the end such as "while ensuring that security and privacy requirements are fulfilled" or "taking into account security and privacy issues" [2]. Moreover, across the full (86-page) document discussing various definitions of IoT, the term "human" is only used in the following instances:

- Reduce or avoid the need for human intervention;
- Satisfy human defined needs;
- Sensing human movement and health;
- Making connections between humans and human to things;
- Human control of these "things" from anywhere.

Nonetheless, IoT technology has a clear impact for the people using it. It offers the promise of smarter factories, greener and safer homes, healthier self, saving of resources, as well as improved cities and infrastructure. At the same time, these benefits come alongside a new wave of privacy and security concerns which are closely related to the human aspects of the technology. Recognition of the human dimension in cyber security is well established, with acknowledgment that people can be barriers to creating effective technology and that there is a need to focus on user support and usable solutions [3]. Therefore, in this connected world of IoT, it must be ensured that cybersecurity strategies do not only include technical solutions but must include a focus on human behavior and security culture.

Human factors pose a recognized security risk to organizations, but many businesses offer strategies to mitigate these risks, such as skills training, employee governance, and business processes overseen by security professionals. However,

IoT devices in industry have made the move from organizations and workplaces to the homes of individuals where users will have responsibility for their own devices rather than there being an IT department to oversee them and additional layers of technology to protect them.

In many cases, these personal IoT devices seem more benign than on a factory floor as they do not seem to control major processes of an individual's daily lives, only devices such as speakers and fridges. However, if these consumer-level devices are not treated securely, they can cause major harm in the hands of a malicious individuals [4]. IoT serves to put far more security and privacy-relevant technology into people's homes, without them necessarily recognizing it as technology that they need to protect and manage in the same manner as their traditional computing devices. In many cases, we have a community of users that are already lacking in their ability to understand and protect against threats on their standard devices, who are now taking the opportunity to introduce more technology, readily accepting the benefits, but unaware that they are also implicitly being dealt a responsibility to secure it.

This chapter gives a specific focus to IoT devices in a domestic scenario, as this represents the context in which the human aspect is most pronounced. Not only does it involve technologies and data that are most personal to the individual, but it also deals with a user community that cannot be relied upon to be have any specific expertise in cybersecurity (nor indeed to be more broadly technology-literate, beyond the ability to operate it at the user level). The discussion begins by looking further into the nature of IoT devices and interactions within the user's home environment, which serves to highlight the breadth of functionality and data that ultimately require protection against compromise. Attention is then given to the nature of security issues of such devices, and more particularly, the related impact for their users. This leads into the main focus of the discussion, which is an examination of the IoT security and privacy challenges from the human factors perspective, including the trade-offs that users may be expected to make in terms of providing and sharing their data, as well as the unexpected (and potentially unwelcome) burden that may be faced as a result of the volume and range of smart devices that they find themselves using. The discussion concludes with a look toward what could be done to improve the situation from the user's perspective, bearing in mind the things that they need to achieve and how appropriate presentation of the technologies may assist them in doing so.

2.2 An Overview of the Domestic IoT Environment

The prospect of all electronic devices in the home and workplace forming part of a single, interconnected network has the potential to affect everyone and every business [1], where the IoT population is expected to outnumber the human population tenfold by 2025. Figures from IHS Markit [5] suggest that there were over

27 billion IoT devices back in 2017, with the volume being forecast to more than quadruple – to 125 billion – by 2030 (with an annual growth averaging 12%). In terms of the devices contributing to this growth, various online sites posted their top selling IoT devices for 2019, which includes the following types [6]:

- *Smart appliances.* This includes smart phones, tablets, smart TVs, and gaming consoles;
- *Home voice controllers.* These devices allow for the controlling of media, alarms, lights, thermostats, etc., via voice commands;
- *Security devices.* This includes smart doorbells, home cameras, locks, alarms, and systems which allow the user to answer a door remotely or see what is going on at the home;
- *Home automation devices.* This includes lights, thermostats, plugs, and air quality sensors that allow for automatic sensing and controlling of the environment. Users can manage and monitor the device from any location hassle-free.

Figure 2.1 depicts the range of connected, smart devices that may now be found in the domestic IoT context. Looking at the items on show, we have traditional IT devices such as desktop computers, laptops, and printers, as well as their somewhat more recent relations such as smartphones and tablets (which users will still typically regard as "computing" devices). However, the typical domestic environment now extends far beyond these. For example, we also have some things that were previously viewed as "technology" but not necessarily smart or connected, such as televisions, gaming consoles, and cameras. Moreover, we also have a further range of items that we have not previously viewed as digital technologies in this sense at all (such as speakers, toys, home appliances such as fridges, wristwatches, and light bulbs), where all of these things now have the potential to be online and include some element of smart functionality. It should also be noted that there are numerous other options that are not included in this stylised depiction, such as heating systems, and a full range of potential appliances (e.g. in addition to the fridge, we can also have smart washing machines, vacuums, kettles, coffee machines, and various others). This already highlights the sheer range of devices that a home user may now have around them, and the point is amplified when we consider that there will then be multiple instances of many of the device types. Indeed, even now when the IoT is essentially still in its infancy, it would not be atypical to find tens of connected devices within the same household.

Of course, in a domestic context, each of the devices will generally be something that people have elected to have around them (although, as later discussion will highlight, they will not necessarily have done so with a specific wish or intention to have a *smart* device). As such, we can generally assume that each device is perceived to bring some sort of benefit. At the same time, however, the fact remains

Figure 2.1 Domestic, smart Internet of Things devices which can be found in a typical home environment.

that these are smart devices that bring with it considerations that have not yet been applied to the previous generations of technology. All will to some extent be storing and sharing data about the users and their environment, which brings with it associated issues around security and privacy of the information. In addition, the security dimension includes ensuring that the devices are free from other forms of compromise – for example disrupting their operation or availability – given that they are performing functions that are necessary to the household concerned.

So while all the devices are there for people, they also represent a growing challenge from the perspective of technology management. There are now more connected devices in the home than we would have seen in small organizations just a few years earlier. The resulting impact equation is simple: more devices mean more to manage. In parallel, it also becomes harder for the average user to understand, track, and control. Indeed, we are already at a stage where many users would struggle to recall all the devices in their IoT environment, let alone to accurately comprehend the data being stored on and transferred by them.

In practice, the connectivity of these devices can differ, but a variety of technologies and protocols are required when utilizing IoT devices in a home, including LTE, Wi-Fi, and Bluetooth. Figure 2.2 indicates how a human in a smart home can possibly interact with various IoT devices and how these various IoT devices are interlinked with each other.

It can be seen from the figure that the user has a smartphone which is connected to the Internet via cellular technology as well as through Wi-Fi via a home router. In addition to the smartphone being connected to the Internet, the user utilizes the smartphone to manage various smart devices, which include smart home sensors and controllers which connect to the use via one single technology (such as Wi-Fi), dual-mode smart home sensors and controllers, which can be controlled via Wi-Fi as well as Bluetooth directly from the smartphone, as well as multimode smart home sensors and controllers where voice control can be integrated as well. It can be seen from the figure that the user interacts with these devices using a range of technologies, as well as how these devices are inherently connected to each other, forming a complex network in the home.

In addition to IoT devices for the home, the growth of wearable and personal IoT devices such as smart watches and fitness trackers have seen exponential growth. These devices generally have the following features:

- Link to the user's smartphone for notifications, synchronisation via an app;
- Incorporate GPS tracking;
- Incorporate health tracking, such as heart rate monitoring, sleep tracking, and activity measurements.

The largest difference between home IoT devices and these wearables are that these devices do not remain stationary and they move around with an individual, moving from network to network as the individual goes about his day. This means that these personal devices not only connect to the home network of the individual, but these devices are also exposed to unknown foreign networks as well. Building upon the earlier diagram, Figure 2.3 illustrates that the addition of wearable devices makes the connectivity and technology requirements of an individual much more complex.

Figure 2.3 shows that the utilization of a smart wearable device connects to the user's smartphone, can be controlled by the user through voice commands, and connects to the Internet via cellular technology. When compared to Figure 2.2, it can be seen that the addition of a wearable device adds not only another level of complexity to the home network but also to any network which the user might utilize outside of the home as the wearable device moves with the user.

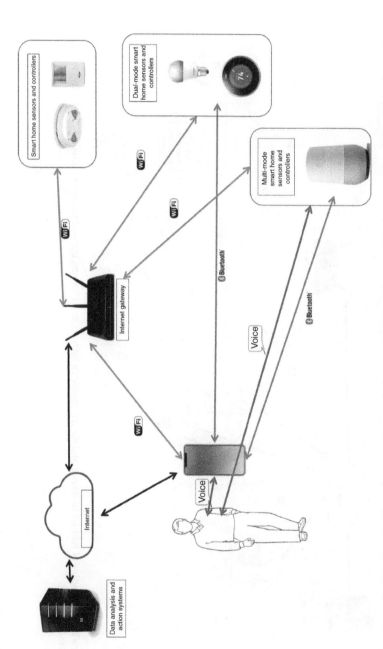

Figure 2.2 Examples of human-to-device and device-to-device interactions showing how humans and devices are interlinked with each other.

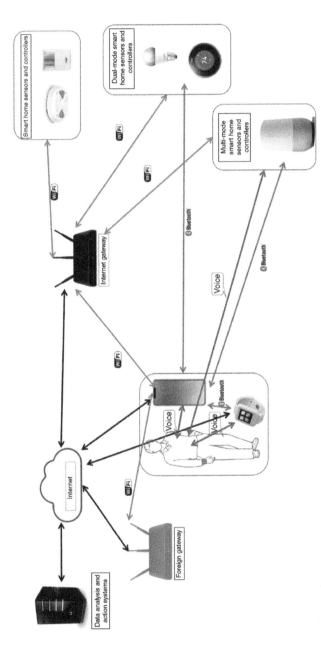

Figure 2.3 Examples of human-to-device and device-to-device interactions showing the additional interlinking when using wearable devices.

These personal and home devices are produced to appeal to individuals who are technically aware and enjoy the use of technology. However, security and privacy may not be the highest priorities of many of these manufacturers, where connectivity and function are typically considered more important in appealing to customers. The ever-growing number of IoT devices, the diversity of these devices, their different capabilities, various parties handling data as well as the countless areas and ways they can be deployed, make security and privacy a unique challenge [7]. In addition, with these personal and home devices, the users are responsible for their own devices and the security surrounding these devices. When these devices are not treated securely, they can cause major personal harm in the hands of a malicious individuals.

2.3 Security Issues and the IoT Landscape

The last few years have seen a significant rise in IoT-centric attacks, many of which have exploited vulnerabilities in the way they have been supplied or configured. These have included attacks that may directly affect the user or their family, as well as attacks that may use their devices to affect others. Headline grabbing examples of the former include incidents demonstrating that devices such as children's toys and baby monitors were vulnerable to attack [8], whereas the hijack of user devices was notably illustrated by the Mirai botnet, which used compromised IoT devices to launch Distributed Denial of Service attacks [9].

As a matter of principle, developers of IoT devices and systems have a collective obligation to ensure they do not expose users to potential harm. However, IoT is a typical example of where technology leads and law and policy follow. As with other technologies in the past, such as wireless access points, the recognition of security aspects tended to follow on some time after a volume of users had already adopted them. As such, security in IoT devices has not been perceived to be a consideration from the outset, leading to potential misconceptions on the part of users. Guidelines and regulations are emerging that set some expectations on security from the vendor/developer side, but not all issues are yet addressed across the board. Many companies go to great lengths to incorporate security into the design phase of IoT devices and services they sell globally [10], but not all of them. The literature cites various security concerns relating to the development and design of the technology itself, including insecure data transfer and storage, use of insecure or outdated components, insecure default settings, and lack of secure update mechanisms [11, 12]. These security issues can lead to a range of IoT-related attacks, which users might not have a lot of say in. Research suggests that the most common technical IoT-related attacks include malware exploitation, man-in-the-middle attacks, privilege escalation, eavesdropping, brute-force password attacks, and Denial of Service attacks [13, 14].

IoT technology has increased the number and magnitude of attack surfaces in a network, which includes third-party IT infrastructure, consumers, and brokers as well as interwoven networks of sensing and actuating devices and servers providing processing, applications, storage, and analytics [15]. This complex environment with a multitude of attack surfaces makes security and privacy difficult to address as there is no one-size-fits-all solution, and every poorly secured device that is connected online has the potential to affect the security and resilience of the network [16].

It is, however, essential for all to understand that challenges on a human level are equally prevalent. Users of personal and home IoT devices must understand that the security of these devices is mainly the responsibility of the individual owning and operating these devices. Individuals continue to utilize and expand on these devices as they believe that the benefits of IoT solutions outweigh any perceived risks that they might face [17]. However, when these devices are not secured and securely used it can lead to serious security and privacy issues for individuals.

Privacy and security go hand in hand when considering the human. When sharing information with other people, a relationship comes down to three factors:

1. We rely on our ability to only disclose information we are comfortable sharing with another individual;
2. We trust the other individual to keep the shared data private and not to use it in ways that can cause harm or conflict with our interests;
3. When entrusting another individual with information, we trust that they will use and store it in a secure way and will not disclose it to unauthorized parties.

In the context of IoT, the trust relationship between the user and the device vendor remains the same: We trust third parties not to abuse the data generated by our IoT devices, we rely on the ability to control the collection and use of that data, and we rely on the vendors to keep our data secure so that it cannot be abused by unauthorized individuals [18].

Therefore, in the ideal IoT environment, users will have the ability to do the following [18]:

- Customize the data collection of IoT devices in an easy way. IoT devices and their applications should enable the user to find out what information is collected and shared and enable this to be customized.
- Control how the user's information collected by their IoT devices is shared and determine who has access to the data from devices. Users should be able to blind and mute devices and to have a say in how IoT data is analyzed or shared with third parties.
- Determine how identifiable one is when undertaking online or offline activities and the ability to control one's digital footprint. Users should understand where information about them has gone and how long it is kept and have the option to delete any information collected by their devices.

- Determine the overall security of IoT devices and its surrounding network environment. Users should be able to view the security status of their environment and the ability to improve it if required.

Privacy can be defined as "support(ing) and empower(ing) people with the option to withdraw from the gaze of, and interactions with, others at will, and the right to respect for their personal space, to create solitude and reserve from others" [19]. When considering privacy in the light of IoT devices, users must be able to enjoy the benefits of IoT devices without privacy risk, or to be able to make informed choices relating to the associated risks [18].

2.4 Human Factors Challenges in IoT Security and Privacy

As previously indicated, the human factor is often cited as one of the biggest challenges facing effective information security. However, this should not be taken to mean that people are necessarily at fault for the issues, or indeed that they are acting in a deliberate or malicious manner. Indeed, a variety of reasons can explain inadvertent human errors and unintentional actions in businesses and organizations which can lead to security issues, including [3]:

- Security protocols and systems which might be too confusing or too difficult for individuals to engage in effectively;
- A lack of understanding by individuals about the importance of the data, software, and systems within an organization;
- Ignorance about the level of risk attached to the assets for which individuals have direct responsibility for;
- A lack of understanding about how an individual's behavior could be putting the business assets at risk.

When considering the use of home and personal IoT devices, the reasons for concern can be expanded to include the following:

- A lack of understanding of the technology itself, including the setup, deployment, and maintenance as it is completely in the hands of the user with no IT department to assist;
- A lack of knowledge relating to the various levels of security design of different devices, where users may focus on functionality instead of security;
- Individuals do not read privacy policies for every device or app they buy. Moreover, even if they attempt to do so, most are typically written in legal language unintelligible to the average consumer [20];

- Lack of knowledge relating to the configuration of devices where unnecessary/unused functions can be disabled but left on default settings. Also, a variety of devices require multiple, inconsistent management/configuration interfaces.

As privacy requires a relationship between parties, the lack of knowledge from the users' side relating to what they share, to whom they are sharing with and how that information is used can lead to significant security and privacy concerns. Indeed, while users readily use smart devices, they may have very little conception of the extent to which their devices are interconnected. Many do not significantly consider what they might be sharing in the process of using them, or their perceptions are misaligned to the reality [21]. Moreover, many have a false sense of anonymity, overlooking the fact that the device is already "logged in" under *someone's* account, and all interactions are being associated with it. If their personal information sharing is directly highlighted (along with the potential for capturing behavioral analytics), then users often express concern [22], but remain with few actionable options (with only a coarse, high-level view of what is going on), and consequently do not feel informed or in control [23].

These points all relate to a user's lack of understanding or knowledge relating to either the technology itself or the security implications relating to insecure setup. In addition to these points, there exist other factors which must be considered when talking about human factors in IoT security. They include the following:

- *Freedom of choice.* When using a device, does an individual have a choice to be recorded/tracked by a device? Does the device provide options to the user to control their data and its uses?
- *Privacy issues.* Data of many devices travel through many communication layers and third-party services, with limited or no corporate transparency. Are the individual's privacy choices respected and implemented throughout the chain of information flow?
- *Security issues.* Are the home and wearable IoT devices, as well as the related data services, secure from vulnerabilities? With the complexities introduced by IoT devices in a network, one poorly secured device can potentially affect a user's complete network.

With these points in mind, the following sub-sections present some further consideration of the issues that individuals may be facing from different perspectives of their IoT device usage.

2.4.1 Security Trade-offs for Individuals in IoT

Considering the human factors at play in the use of IoT devices, it can be seen that some appear to sit somewhat in contrast with each other. Some of these trade-offs include the following:

- *Use of IoT devices and privacy.* Consumers are surrendering their privacy when using devices which collect their personal data sometimes without their knowledge [24, 25];
- *Security and comfort.* Home automation devices, such as smart thermostats, fridges, and cameras, can ease daily operations in the home, but provide malicious actors with opportunities to control factors in a user's home [26];
- *Freedom of choice and ease of use.* Users might be presented with some freedom of choice relating to the privacy setup of a device, which is contrasted with ease of use of the device.

These contrasting concepts are further discussed below to highlight the complexities at play when using IoT devices. The use of these devices provides a myriad of choices to individuals, which clearly can lead to security issues to fall by the wayside.

One of the most dangerous factors relate to the collection of private data by devices. In many cases, users might surrender private information willingly to assist with certain daily operations, such as traffic navigation or health tracking. In other cases, however, users do not realize what private data is collected by these devices. Many users simply upgrade their "dumb" devices with a similar device on the market, not understanding the implications of it now being "smart." These new devices now have the capability to track users and collect information, where it most likely does not occur to the users that this is happening. These devices may enable the user to personalise the behavior of smart objects and associated applications, but it may not be that simple for the average consumer, especially if the device was not bought with the use of the smart capabilities in mind. In many cases, features are enabled by default and a user with limited knowledge of technology might not know how to personally configure these devices or may not even be aware that it is an issue to know about. New smart devices are sold as "consumer electronics" but is raising the bar in terms of the technical knowledge needed to own and operate it effectively. Previous generations of consumer electronics (e.g. audio and video equipment) offered the opportunity for more informed consumers to get into the technical details, where the average user could just turn on and use the device without this level of concern.

The "always on" nature of some devices, which integrates features such as voice recognition or vision features, are providing additional privacy concerns as they continuously listen to conversations or watch for activity and selectively transmit this data to cloud services for processing. In 2015, the privacy policy of Samsung Smart TV's came under fire as the small print in the privacy policy stated that its Smart TV's voice recognition system can capture your private conversations and might pass them onto third parties [24, 27]. The policy stated that the company may "capture voice commands and associated texts so that we can provide you with Voice Recognition features and evaluate and improve the features." In addition, the policy stated the following: "Please be aware that if your spoken words include personal or other sensitive information, that information will be among the data captured and transmitted to a third party through your use of Voice Recognition." This incident caused Samsung to edit its privacy policy and clarify the Smart TV's data collection practices stating that "the voice data consists of TV commands, or search sentences only," and that "industry-standard security safeguards and practices" are implemented [27]. The collection of this type of information exposes legal and regulatory challenges facing data protection and privacy law. There exist many such examples not only limited to home devices but also includes wearable devices, smart phones, and especially smartphone apps.

The laws about data privacy are very clear in providing a guideline to data collecting companies. They must collect and use the data for specific purposes only after obtaining the consent of the concerned person. However, individuals generally do not familiarize themselves with the privacy policies for every device or app they buy. When unpacking a new device, the average nontechnical individual is likely to prioritize function over security, which means that they will select "I agree" on all the provided setup screens just to get it working. The difficulty of understanding these policies and not prioritising the content then leads to cases where individuals surrender their privacy even without them knowing the full extent of the disclosure.

2.4.2 Data Ownership and Use

The collection of data via IoT devices brings further questions relating to the ownership of data. When considering IoT devices, several players are involved in the journey of the data, which can include the user, hardware manufacturer, app developer, software business who processes the data, provider of database architecture, and other possible third parties [28, 29]. The data generated by IoT devices are hugely valuable and all players in the IoT data chain may want to claim rights over it.

Many law experts seem to argue that the "maker" is the first owner of the data, which can be seen as the entity who "made the commercial decision to collect the data and made the investment in carrying it out." This suggests that entities toward the top of the chain are more likely to be the maker [28, 29]. Data transfer between parties also does not imply ownership transfer, meaning that the data owner may allow third parties to access their data, but that some contractual agreement must ensure that if the owner asks the third party using its data to delete it, it should be done [28]. Considering these arguments, it is the position of these experts that end users utilizing off-the-shelf IoT devices do not have ownership rights to the data gathered by these devices [28, 29]. This does not mean that these companies own the user's personal data, but that they own the nonpersonal machine-generated data collected by these devices. This means that a user cannot claim ownership of the details relating to TV usage patterns or when lights were turned on and off.

Information relating to TV usage or lights might not seem too important to users, but when sensitive data such as health data is considered, this can become a greater concern to users. Experts in the field agree that in most instances, medical records are not owned by the patient [30, 31]. As the health care provider owns the media in which the information is recorded and stored, it gains the property right of possession of data and becomes the legal custodian of an individual's health care record [30], where patients only have the specific legal rights to access their information. There exist many good arguments for opening health records to third parties to improve health care for individuals and populations, including the enrichment of data on symptoms, diseases, diagnosis, treatments, and prescriptions. Advantages can also include better patient care due to easier access to information by different professionals [30].

However, the collection, handling, and storage of large amounts of data bring challenges relating to privacy. In a traditional setting, the individual formed an accepted social contract with the healthcare system sharing health information with medical personnel directly, implying rights as well as responsibilities for consent, protection, and privacy. In this new context, health data becomes synonymous with consumer data, where the rights of the data as well as technical security measures are not always clearly communicated or disclosed and that currently patients have little say in their own health care data [30]. There exist many cases where patients struggle to obtain data generated from their medical devices, such as pacemakers, where data are easily accessible and available to healthcare professionals but not to them personally [32].

The case relating to data collected via health applications, smart watches, and fitness trackers seem even more complicated. Legal environments differ from country to country, but this personal health data are mainly subjected to industry-defined terms of conditions, as the case with other home IoT devices [30]. Data generated by these devices are not considered part of your health record data as generated by healthcare professionals [33]. This means that users' sensitive health data can be used for personalized online marketing, internal research, and sale to third parties without providing users with opt-out clauses or options to delete their data. Many fitness trackers do not provide users the capability to obtain their detailed fitness data, with others only providing users with summaries or high-level information [34]. In addition, users are given free fitness devices by their health insurance, with the requirement that they log a certain level of activity per day, enabling these companies to gain even more information on these individuals. Many of the articles dedicated to these issues advise individuals to ensure that they understand their rights relating to private health records. They are advised to consider the privacy and data terms and conditions when signing up for certain services or devices. However, as with personal IoT devices, most policies are unintelligible to the average consumer or, although provided with a "choice," cannot really deny the conditions as they need to use this product or service.

2.4.3 Device Management and Administration Responsibilities

The use of home or personal IoT devices places the responsibility solely on the user to ensure secure management of the devices. Vendors can issue updates to firmware and software, but unless the process is entirely automated the responsibility for installing the upgrade ultimately still lies with the user. The first issue regarding device upgrades relates to the manufacturer providing secure up-to-date updates. Manufacturers may design devices with the inability for firmware updates or stop the support for devices which are still in use, compromising the future security of the device. This means that devices become vulnerable to attacks when they are not kept up to date with the latest security updates. Not running updates can lead to security breaches of not only the compromised device but also the user's network and the systems of the companies that manufacture them [12, 35, 36]. In addition, many shipped IoT devices come with a default username and passwords set by the manufacturer.

This issue has been recognized by US Senate Bill 327 on the information privacy of connected devices. Specifically, this introduced a requirement that, from the start of 2020, manufacturers of Internet-connected devices must:

> *equip the device with a reasonable security feature or features that are appropriate to the nature and function of the device, appropriate to the information it may collect, contain, or transmit, and designed to protect the device and any information contained therein from unauthorized access, destruction, use, modification, or disclosure* [37].

The scope of the Bill encompasses all connected devices rather than specifically IoT devices, but clearly has particular relevance in this context. In practical terms, one of the underlying impacts is that default passwords must now be unique to each device (i.e. rather than all devices shipping with a universal factory default).

In a similar move, the UK government's Department for Digital, Culture, Media and Sport and the National Cyber Security Centre have jointly issued a code of practice for the security of consumer IoT devices [38]. This proposes thirteen guidelines, covering various aspects of protection, including elimination of universal defaults, storage and communication of personal data, and the ability to keep the devices updated. The hope is for device developers, manufacturers, and retailers to voluntarily adopt practices that help to protect those using them.

In this situation, users are advised to ensure that they only use devices from vendors who will possibly be around for years to come and have proven to continuously provide updates to products when there is an issue [12]. Users are also advised to change the default usernames and passwords as soon as they start to use the device. This can become complicated as the average user might not know how to do these checks or change the device's user credentials.

Second, users are required to apply firmware updates, software updates, and security patches that run on their IoT devices as well as keep track of the versions that are deployed on each device. An individual must keep track of which updates are available and apply them consistently across all their devices which may run on a variety of platforms. Not all vendors inform users of available updates, which means that users must constantly check for updates for devices which do not automatically provide communication relating to an update. In addition, not all IoT devices support over-the-air updates, so some IoT devices might need to be physically accessed via network cable to apply updates via a computer [35].

Devices which run from similar platforms can be managed through a device manager which can push updates automatically to devices and help ensure that only legitimate updates are applied to devices. This can be helpful to users if they use similar platforms, but with the range of IoT devices available, not all devices might be supported by a single platform. Users end up with multiple, inconsistent management, and configuration interfaces which can complicate the management of these devices. With all these devices and platforms, many devices may not be updated regularly compromising the device as well as the user network.

2.4.4 The Age of Unwanted Intelligence

All the preceding discussions on trading data for functionality and the additional responsibilities in managing and maintaining devices have thus far overlooked a potentially significant point – that users did not actually *ask* for this in the first place. As mentioned, many have upgraded from dumb devices into the new realm of smart technology – but they were not necessarily looking to do so. They may simply have discovered that they had no other option, as there are fewer and fewer dumb devices available in the market. For example, it is now increasingly difficult to buy a new television that does not have Smart TV functionality (particularly if you want certain other features, such as large screen and high resolution). Moreover, other devices tend to be following a similar path, and owners can find themselves in the situation where they have to connect the device to a network to get it to work at all (or to stop it complaining that it is not connected). So we have moved from a position where people could simply buy a device, power it up, and use the functionality they were seeking, to a situation where the same basic functionality is only available in return for network connectivity and registration of personal details. Of course, many of the new features can be beneficial and even reluctant users can ultimately find themselves having a use for them. But the situation remains that users can ultimately feel forced and even coerced into using technology features that they were not looking for and being landed with associated security and privacy concerns as a result.

To consider what this may mean, we can stay with the example of Smart TVs. One of the consequences of the smart capability in this context is the potential to run apps, and of course, as we have seen repeatedly in other contexts, the ability to install software immediately offers an opportunity for the spread of malware. As a result, users of potential target devices now find themselves needing to be aware of the threat and take safeguards against it. This point is well illustrated by the following example from 2019 of Samsung Smart TV users being advised to perform regular scans for malware. Of course, regular users of technology would have been very familiar with at least the notion of malware by this point, with the threat having been around on traditional computers since the early 1990s. However, finding that it was now apparently a consideration on their TV was perhaps an unwelcome novelty. Nonetheless, this was exactly the situation that Samsung presented to its customers, with the following being an exact quote from a related Tweet sent by Samsung Support [39]:

> *Scanning your computer for malware viruses is important to keep it running smoothly. This is also true for your QLED TV if it's connected to Wi-Fi! Prevent malicious attacks on your TV by scanning for viruses on your TV every few weeks.*

So let us now consider this from the perspective of the user who did not want the smart aspect of the TV in the first place. They have gone from having a device for which the main threat was previously physical theft to having something that they are now being told is at risk of "malicious attack." Moreover, this now introduces an additional overhead for them, of being expected to scan the device on a regular basis (whereas their previous TV maintenance regime may have extended no further than occasionally dusting it off).

To put the warning in some sort of context, the volume of malware actually targeting TVs was very small, and so it was not as if Samsung's customers (or users of other smart TV brands) were at significant risk. However, it would also be fair to observe that the warning is only likely to become more relevant as time goes on. Indeed, we have already seen that this evolution occur with malware on mobile devices. Back in the early 2000s, malware was starting to appear as proof-of-concept on the mobile device platforms of the era. However, much of the commentary on the threat tended to dismiss it as a nonissue because the population of target devices was not an attractive alternative to the established base of potential victims on Windows-based devices. However, within a decade things had transformed, with the arrival of smartphones and particularly with the use of the Android operating system (where the open nature of the app marketplace allowed malware to find its way onto devices without any barriers in its way). As time went on, Android became the dominant platform for malware on mobile devices in a very similar manner to Windows in the PC market. Today, no-one would dismiss malware on mobile devices as mere hype, and users on affected platforms are well advised to be running anti-malware protection as a safeguard. If smart TVs were to follow a similar path, with the devices (or their users) being seen as a viable and valuable target, then the criminal community will come to use malware to target them more significantly.

2.5 Toward Improved User-facing Security in the IoT

Recognizing that smart devices are only going to proliferate further, a key consideration needs to be how to ensure that they are manageable from the user perspective. Unfortunately, but perhaps unsurprisingly, there will not be a one-size-fits-all approach, and it will require more than a single action. This does not make it an impossible goal, but it needs to be recognized as requiring action from several levels and perspectives, with a shared responsibility for both providers and users to play their part.

As a foundation, there are various aspects of security awareness that users will need to have in order to follow good practice. This includes buying from reputable vendors, changing default account access settings, installing updates, and not reusing passwords across multiple devices. These are by no means unique to the IoT context and ought to be part of an overall picture of cybersecurity literacy [40]. However, users must assume increased importance in this situation because it otherwise leaves an increased level of exposure for the user, their environment, and their data. However, while this foundation is important, it is clearly not going to be sufficient. Unlike the organisational context, where users can typically rely upon things being taken care of for them where training and support are available, securing home devices is ultimately left up to the individual, and they are operating without any requirement to show that they are security literate or aware.

Many users struggle to understand or express their requirements in conventional terms and via existing controls. The task of establishing and maintaining a secure home IoT environment could be burdensome – even more so for the average individual who wants to utilize these technologies, but might not be an expert in the field of IoT and security. In these cases, technology can be utilized to handle it and ease the security management of the individual. As such, it is also relevant to consider technical solutions that can assist the home IoT user to secure their devices and safeguard their data.

One such technology is the Databox project (see www.horizon.ac.uk/project/databox/), which focuses on user-controlled privacy. The vision of the Databox is to function as a personal networked device that collates, curates, and mediates access to a user's personal data. The Databox then provides a personal data processing ecosystem which will enable individuals to control the use of their personal data. Looking more widely, while techniques have been proposed to empower users' data ownership [41], they still require a means to take control in a comprehensible and useable manner, rather than facing the heterogeneity and complexity of current privacy options, which can currently be a significant deterrent to effective use. It is necessary to determine appropriate means of conveying feedback around data-sharing behaviors. This includes means of enabling users to understand what type(s) of data are shared in relation to which types of interaction and means of regulating the sharing to enable privacy management and reduce risks such as oversensing by devices within their environment [42].

Linked to this, there would also be value in establishing a user-friendly means of *capturing* and expressing users' data privacy and control requirements and then automatically *propagating* these across different devices and services. This would deliver a more usable means of configuring privacy settings (translating user needs into settings by allowing them to control of how sensitive data will be used) and a frictionless means of applying these across multiple devices/environments in order to reduce both the administrative overhead and the potential for inconsistent data/privacy handling.

It is also relevant to recognize that it is not just about enabling the things that the user wants to do – there is also likely to be an ongoing need to get them to engage with certain management and administration tasks that cannot be entirely automated. Again, there is no single answer as to how to ensure such engagement, but consideration can perhaps be given to how to relate to people on a more human level. For example, given the number of devices within the environment as a whole, there could also be advantages to presenting users with a broader, meta-level view of the security and protection status of their IoT estate. If users have a means of conceptualizing the overall "health" of their environment (e.g. via some form of simplified top-level indication and underlying dashboard that summarizes the protection and patch status on specific devices), then this could potentially be used to engender some form of "Tamagotchi effect" [43] in terms of motivating a desire to take care of things that still require their personal intervention. For example, if people are told that half of the devices in their smart home are "unhappy" (i.e. requiring some form of security-related attention), this may be sufficient to incentivize caring users to take action to "look after" them. Equally, anyone that then fails to act will still realize that they were given a meaningful opportunity to do so if things go wrong later, thus again relating to them at the human level (in this case, basically saying "don't say you weren't warned!").

2.6 Conclusion

In conclusion, it is clear that the security and privacy aspects of IoT technologies are inextricably linked to the needs and understanding of the people that use them. It is not just a question of people needing to understand that there are issues, nor simply a case of needing to use technology-based controls to reduce risks. As with other strands of IT, the most effective means of addressing cyber security lies in the use of an appropriate combination of technology and human-focused safeguards. One of the key differences with IoT devices – particularly in domestic context – is that they both amplify the task that needs to be addressed and do so among a user community that is not looking to acquire new security responsibilities.

While this chapter has not set out to present definitive solutions, it has laid out a number of challenges that need to be considered and highlighted the ways in which these human-facing issues require similarly human-focused approaches to address them. In this sense, it provides a reference point for those designing and deploying devices, reminding them of the need to support users in achieving and managing their security and privacy with the same diligence as they support the core or primary functionality of the device.

Acknowledgments

Icons made by Pixel perfect from www.flaticon.com.

References

1 EY (2015). Cybersecurity and the Internet of Things. *EY*. https://www.ey.com/Publication/vwLUAssets/EY-cybersecurity-and-the-internet-of-things/$FILE/EY-cybersecurity-and-the-internet-of-things.pdf (accessed March 2015).

2 Minerva, R., Biru, A. and Rotondi, D. (2015). Towards a definition of the Internet of Things (IoT). *Rev 1*. https://iot.ieee.org/images/files/pdf/IEEE_IoT_Towards_Definition_Internet_of_Things_Revision1_27MAY15.pdf (accessed May 2015).

3 Hadlington, L. (2018). The "human factor" in cybersecurity: exploring the accidental insider. In: *Psychological and Behavioral Examinations in Cyber Security* (eds. J. McAlaney et al.), 46–63. IGI Global.

4 Kilpatrick, H. (2018). Cybersecurity challenges could make or break IoT. *IoT for All*. https://www.iotforall.com/iot-cybersecurity-challenges/ (accessed September 2018).

5 IHS Markit (2017). The Internet of Things: a movement, not a market. *IHS Markit e-Book*. https://cdn.ihs.com/www/pdf/IoT_ebook.pdf.

6 Kenneth, K. (2019). Top 10 most popular IoT devices In 2020. *Robots.net*. https://robots.net/tech-reviews/top-iot-devices/ (accessed November 2019).

7 Feeney, M. (2019). How to manage Internet of Things (IoT) security in 2020. *AT&T Business*. https://cybersecurity.att.com/blogs/security-essentials/how-to-manage-internet-of-things-iot-security (accessed September 2019).

8 Laughlin, A. (2017). Safety alert: see how easy it is for almost anyone to hack your child's connected toys. https://www.which.co.uk/news/2017/11/safety-alert-see-how-easy-it-is-for-almost-anyone-to-hack-your-childs-connected-toys/ (accessed 14 November).

9 Williams, C. (2018). IoT gadgets flooded DNS biz Dyn to take down big name websites. *The Register*. https://www.theregister.co.uk/2016/10/21/dyn_dns_ddos_explained/ (accessed 21 October 2018).

10 Eggers, M.J. (2017). IoT cyber policy. *NIST IoT Cybersecurity Colloquium*. https://www.nist.gov/system/files/documents/2017/10/23/mattheweggers_slides.pdf (accessed October 2017).

11 Paul, F. (2019). Top 10 IoT vulnerabilities. *Network World*. https://www.networkworld.com/article/3332032/top-10-iot-vulnerabilities.html (accessed January 2019).

12 Young Entrepreneur Council (2018). 10 big security concerns about IoT for business (and how to protect yourself). *Forbes*. https://www.forbes.com/sites/theyec/2018/07/31/10-big-security-concerns-about-iot-for-business-and-how-to-protect-yourself/#2830ad5b7416 (accessed July 2018).

13 IoT Analytics (2017). IoT security market report 2017 – 2020. *IoT Analytics* (September 2017).

14 Nedbal, M. (2019). IoT insecurity: 6 common attacks and how to protect customers. *Channel Partners*. https://www.channelpartnersonline.com/blog/iot-insecurity-6-common-attacks-and-how-to-protect-customers/ (accessed 2019).

15 Krautkremer, T. (2014). The Internet of Things needs a network of clouds. *Silicon Angle.* https://siliconangle.com/2014/07/03/the-internet-of-things-needs-a-network-of-clouds/ (accessed July 2014).

16 Iqbal, M.A., Olaleye, O.G., and Bayoumi, M.A. (2016). A review on Internet of Things (Iot): security and privacy requirements and the solution approaches. *Global Journal of Computer Science and Technology: E-Network, Web & Security* 16 (7) https://globaljournals.org/item/6471-a-review-on-internet-of-things-iot-security-and-privacy-requirements-and-the-solution-approaches.

17 Curtis, B. (2019). IoT device security concerns could limit IoT growth. *Forbes.* https://www.forbes.com/sites/moorinsights/2019/11/15/iot-device-security-concerns-could-limit-iot-growth/#5f87f12c6275 (accessed November 2019).

18 Internet Society (2019). Policy brief: IoT privacy for policymakers. *Internet Society.* https://www.internetsociety.org/policybriefs/iot-privacy-for-policymakers/ (accessed September 2019).

19 Warren, S.D. and Brandeis, L.D. (1890). The right to privacy. *Harvard Law Review* 4 (5): 193–220.

20 Furnell, S. and Phippen, A. (2012). Online privacy: a matter of policy? *Computer Fraud & Security* 2012 (8): 12–18. https://www.sciencedirect.com/science/article/pii/S1361372312700830.

21 Al-Ameen, M.N., Chauhan, A., Ahsan, M.A.M. and Kocabas, H. (2020). Most companies share whatever they can to make money!: comparing user's perceptions with the data practices of IoT devices. *Human Aspects of Information Security and Assurance*, IFIP Advances in Information and Communication Technology 593: Springer International Publishing. pp. 329–340.

22 Haney, J., Furman, S., Acar, Y. and Theofanos, M. (2019). Perceptions of smart home privacy and security responsibility, concerns, and mitigations. *15th Symposium on Usable Privacy and Security (SOUPS 2019)*.

23 Tabassum, M., Kosinski, T., and Lipford, H. (2019). "I don't own the data": end user perceptions of smart home device data practices and risks. *Proceedings of the Fifteenth Symposium on Usable Privacy and Security*, Santa Clara, USA 12-13 August 2019, pp. 435–450.

24 Bannan, C. (2016). The IoT threat to privacy. *Tech Crunch.* https://techcrunch.com/2016/08/14/the-iot-threat-to-privacy/ (accessed August 2016).

25 Morrow, S. (2018). The importance of privacy and IoT. *IoT for All.* https://www.iotforall.com/five-reasons-privacy-iot-incompatible/ (accessed October 2018).

26 Banafa, A. (2017). Three major challenges facing IoT. *IEEE Internet of Things.* https://iot.ieee.org/newsletter/march-2017/three-major-challenges-facing-iot.html (accessed March 2017).

27 Matyszczyk, C. (2015). Samsung's warning: Our Smart TVs record your living room chatter. *C Net*. https://www.cnet.com/news/samsungs-warning-our-smart-tvs-record-your-living-room-chatter/ (accessed February 2015).

28 Best, J. (2016). Who really owns your Internet of Things data? *ZD Net*. https://www.zdnet.com/article/who-really-owns-your-internet-of-things-data/ (accessed January 2016).

29 Rendle, A. (2014). Who owns the data in the Internet of Things? *Taylor Wessing*. https://www.taylorwessing.com/download/article_data_lot.html (accessed February 2014).

30 Kostkova, P., Brewer, H., de Lusignan, S. et al. (2016). Who owns the data? Open data for healthcare. *Frontiers in Public Health* 4.

31 Sharma, R. (2018). Who really owns your health data? *Forbes*. https://www.forbes.com/sites/forbestechcouncil/2018/04/23/who-really-owns-your-health-data/#5aaa24b76d62 (accessed April 2018).

32 Alexander, N. (2018). My pacemaker is tracking me from inside my body. *The Atlantic*. https://www.theatlantic.com/technology/archive/2018/01/my-pacemaker-is-tracking-me-from-inside-my-body/551681/ (accessed January 2018).

33 Weinstein, M. (2016). What your fitbit doesn't want you to know. *Huffpost*. https://www.huffpost.com/entry/what-your-fitbit-doesnt-w_b_8851664 (accessed December 2016).

34 Lee, S.M. (2015). Who owns your steps?. *BuzzFeed News*. https://www.buzzfeednews.com/article/stephaniemlee/who-owns-your-steps (accessed July 2015).

35 Gerber, A. and Kansal, S. (2017). Top 10 IoT security challenges. *IBM*. https://developer.ibm.com/technologies/iot/articles/iot-top-10-iot-security-challenges (accessed November 2017).

36 Shah, V. (2019). 9 main security challenges for the future of the Internet of Things (IoT). *readwrite*. https://readwrite.com/2019/09/05/9-main-security-challenges-for-the-future-of-the-internet-of-things-iot (accessed September 2019).

37 California Legislature (2018). SB-327 Information privacy: connected devices. Senate Bill No. 327. Chapter 886. https://leginfo.legislature.ca.gov/faces/billTextClient.xhtml?bill_id=201720180SB327 (accessed 28 September 2018).

38 DCMS (2018). Code of practice for consumer IoT security. *Department for Digital, Culture, Media and Sport*. https://www.gov.uk/government/publications/secure-by-design/code-of-practice-for-consumer-iot-security (accessed 14 October 2018).

39 BBC News (2019). Samsung TVs should be regularly virus-checked, the company says. *BBC News online*. https://www.bbc.co.uk/news/technology-48664251 (accessed 17 June 2019).

40 Furnell, S. and Moore, L. (2014). Security literacy: the missing link in today's online society? *Computer Fraud & Security* 2014 (5): 12–18. https://www .sciencedirect.com/science/article/pii/S1361372314704919.

41 Shafagh, H., Burkhalter, L., Hithnawi, A. and Duquennoy, S. (2017). Towards blockchain-based auditable storage and sharing of IoT data. arXiv.org. https://arxiv.org/abs/1705.08230 (accessed 14 November 2017).

42 Bolton, C., Fu, K., Hester, J., and Han, J. (2020). Inside risks – how to curtail oversensing in the home. *Communications of the ACM* 63 (6): 20–24.

43 Dormehl, L. (2019). The Tamagotchi Effect: How digital pets shaped the tech habits of a generation. *digitaltrends*. https://www.digitaltrends.com/cool-tech/ how-tamagotchi-shaped-tech/ (accessed 29 May 2019).

3

Applying Zero Trust Security Principles to Defence Mechanisms Against Data Exfiltration Attacks

Hugo Egerton[1], Mohammad Hammoudeh[1], Devrim Unal[2], and Bamidele Adebisi[3]

[1] *Department of Computing and Mathematics, Manchester Metropolitan University, Manchester, UK*
[2] *KINDI Center for Computing Research, Qatar University, Doha, Qatar*
[3] *Department of Engineering, Manchester Metropolitan University, Manchester, UK*

3.1 Introduction

Internet-connected devices, including the Internet of Things (IoT) objects, are used for data exfiltration. Data exfiltration is the unauthorized transfer of data from a system by someone with malicious intent or through the use of malware which has been installed on a system [1]. This can be performed both physically and remotely. Remote exfiltration typically utilizes a form of backdoor or malicious program, with the capability of stealthily monitoring any outgoing or incoming traffic and the current data held on the target system [1]. When someone can gain physical access to a target system, then the data exfiltration can be physically performed, such as the transfer of data to an easily concealed external storage drive. The threat of data exfiltration is very prominent as, by the end of Q1 2020, there had been a total of 1,196 publicly disclosed data breaches worldwide, with a total of 8.4 billion records or pieces of data compromised [2]. While many threats of data exfiltration reside on the network level and can be solved with the relevant network-based security deployments, there are also physical threats which require addressing to ensure protection across all areas of a company's infrastructure [3].

Typically, physical threats can be solved with physical measures such as enclosures for systems to prevent tampering, insertion of external storage devices, and the use of physical locks to prevent the theft of a system from the premises [4]. In IoT, due to the deployment of devices in accessible place, physical threats can be far more difficult to address. Additionally, a network-based solution must be designed to monitor and supervise any physical actions performed on any systems

Security and Privacy in the Internet of Things: Architectures, Techniques, and Applications,
First Edition. Edited by Ali Ismail Awad and Jemal Abawajy.

in the network. In this research, we describe such a solution, which is capable of monitoring and preventing unauthorized actions from being performed on systems in the network, such as transfers or modifications of data. Through the use of data fingerprinting and networkwide endpoint rules, any potential data leakage or exfiltration can be mitigated as any data that are considered confidential or any action that is not within the permitted rule list is denied and therefore cannot be performed [5].

As the focus of this mechanism is to prevent data exfiltration attacks from occurring within the internal network, placing the solution against zero-trust security principles is crucial as it assumes that no one can be trusted and everyone, regardless of their clearance or privileges, must have their actions and file use verified.

The zero-trust principle explicitly verifies that the authentication and authorization of all actions are performed regardless of the requesting user's credentials or permissions [6]. While some users will have more privileges than others, they must all be verified with key points such as user account, location, service, and classification of data. This leads to a policy of zero-trust, where both users and administrators alike must go through some form of verification before they can perform actions, especially actions that involve the use of confidential or sensitive data.

Maintaining the least-privileged access zero-trust principle will be achieved by limiting user access rights on each endpoint within the internal network. Using a rule list which is deployed to each endpoint, it stops unauthorized actions from being performed including the use of any system tools, programs, and data. Groups of endpoints can be bound to a specific set of rules, such as segmenting Human Resource (HR) and financial systems so that only those within the financial sector can access finical software and data on a particular set of endpoints. This has the potential to work well in conjunction with an existing deployment such as Active Directory, with it being capable of reinforcing the privileges and access that a user should have to perform specific actions and access data while also blocking any of the prohibited actions from occurring.

Finally, setting policies for the value of data, zero-trust principle will be addressed by the use of data-driven protection, with the sensitivity of data being determined by searching for keywords against all data hosted on the network. By using keyword searches, the level of sensitivity can be determined from how many keywords matches there are so the data can be appropriately placed into two classifications of either nonconfidential or confidential. This can then help in the detection and prevention of any exfiltration attempts using classified information, with file classification being checked during the authorization of any action involving the use of a particular file or number of files. This is also utilized with the endpoint rule list to determine if both the actions can be performed and the corresponding file used in the action contains any confidential information.

This chapter presents the details of a mechanism that is capable of mitigating physical data exfiltration attacks, with a focus on physical vulnerabilities that can be exploited by insiders to acquire unauthorized access to sensitive information. Our key contributions are as follows:

1. A comprehensive review of the various attacks and the associated techniques and technologies they use for data exfiltration.
2. An evaluation of the typical defense mechanisms used to counter or defend against data exfiltration attacks.
3. Develop a new data exfiltration approach and demonstrate various data exfiltration attacks that can be performed, and review the findings.
4. A new method that can effectively counter data exfiltration attacks.

The rest of this chapter is organized as follows: Section 3.2 provides a critical review of recent prominent related literature around data exfiltration and the technologies utilized to perform it with a review of the current academic research. Section 3.3 proposes a defense mechanism that can protect systems against physical methods of data exfiltration attacks. Section 3.4 critically analyzes the proposed defense mechanism and determines its viability as a solution that could be used for real-world scenarios of protecting against data exfiltration attacks. Finally, Section 3.6 concludes and sets future directions.

3.2 Data Exfiltration Types, Attack Mechanisms, and Defence Techniques

The main goal of this section is to explore and provide a full technical understanding of data exfiltration, including attack techniques and defensive mechanisms. As well as examining data exfiltration, more specific areas are investigated, including the preliminaries of insider attacks and approaches to counter data exfiltration.

3.2.1 Types of Data Exfiltration

3.2.1.1 Physical
Physical data exfiltration occurs when a malicious actor can gain physical access to a system or data to perform the attack either by initiating a transfer over the Internet, by physical theft of the device or data, and by manually transferring the data to an external storage device. By having access to the system, it opens up many more opportunities, such as the installation of malware to allow for remote data exfiltration or even the theft of the system.

However, even with physical access to the system, this only becomes an effective means of data exfiltration if the malicious actor can gain access to the system data.

Therefore, if there is any form of password protection or disk encryption on the system, valid credentials must also be acquired so that the system may be accessed and the data made available for transfer.

Other forms of physical media may also be used by malicious actors for data exfiltration, such as printouts of any important documentation and USB drives. Also, with the emergence of bring your own device (BYOD), any of the devices that are brought in by employees can cause an immediate security risk, with devices potentially being at risk of being stolen, lost, or even being used as a entry point by attackers [7].

3.2.1.2 Remote

Remote data exfiltration utilizes a remote connection to allow for the transfer of data from the system without requiring physical access to it. By establishing a remote connection to a system, a malicious actor can not only exfiltrate data from that system but also perform traffic analysis and monitor/intercept incoming and outgoing communications such as e-mails [8]. HTTP/HTTPS, FTP/SFTP, and e-mails are all key examples of exfiltration channels that can be used for remote data exfiltration [9].

For remote data exfiltration attack methods to be viable, malicious actor uses a vulnerability or exploit to access the system from an external network. This can come from a range of areas, but typically, these can stem from outdated software, outdated security updates, undetected malware, or through exploiting any currently running services and protocols on the system. While remote exfiltration might provide more freedom for methods of attack and transferring data, if the vulnerability the malicious actor utilizes to access a system is identified and resolved, they will no longer be able to access the system and must identify an alternative method of access or target of opportunity.

3.2.2 Data Exfiltration Attack Techniques

3.2.2.1 Physical-based
Theft of Device The first attack method for physical data exfiltration is the theft of the target system. This method requires a malicious actor having physical access to both the premises in which the system is located as well as the target itself, which might require some form of identification or clearance before being able to access. This method can be detected very easily and quickly if not done with some planning, especially if the theft occurs during working hours when the device is used consistently.

Devices such as mobile phones and/or tablets are prime targets for this form of data exfiltration as it is much more likely to be identified as a missing device rather than a stolen one, as both can easily be misplaced, especially if employees take part in BYOD [10]. Once stolen, malicious actors can connect the devices to their systems and then use the appropriate tools or techniques to breach into the device and then exfiltrate any stored data such as confidential e-mails and documents [1].

External Storage Device   USB drives and external hard drives can easily be used for data exfiltration, as they simply require being plugged into the target system and then can be used to exfiltrate data from the system. While many companies do engage in protection methods such as disabling USB ports on the systems to protect against this form of data exfiltration, many companies do not do this and leave themselves open to attack.

A report by McAfee found that removable devices accounted for 31% of all data breaches, illustrating just how capable and viable a method this is to exfiltrate data [11]. By copying the data over to an external storage medium, it will leave no trace of clear evidence unless an administrator was to examine the system logs, which further proves that this as an effective attack method. The transference of data from a system to an external medium does not typically require any user intervention once started, so this method can also be left unattended.

Discarded Device Retrieval   When companies or consumers upgrade their infrastructure or systems, they must ensure that they dispose of any unwanted devices correctly. If this is not done, then it opens up the opportunity for malicious actors to be able to retrieve the devices and attempt to exfiltrate any data that is still on the systems.

Companies that do not ensure storage devices such as internal hard drives are wiped and/or removed from systems before being discarded can enable malicious actors to exfiltrate data with minimal effort. Unless an internal hard drive has been formatted several times, data recovery tools can be used on these drives to attempt to recover any residual data that is still on the disk. Even by recovering a small amount of data, it can potentially lead to more severe consequences, especially if the recovered data refer to any credentials or clearance information that can be used by a malicious actor to access confidential information. Personal devices used by everyday consumers of technology are also faced with this threat, as many people simply dispose of their unwanted devices or technology at recycling centers or drop-offs without first ensuring that all data has been erased.

3.2.2.2   Remote-based

Websites   One of the first methods that can be used for remote data exfiltration is through the use of websites to exfiltrate data from a system within an organization to an external network [12]. Websites such as GitHub, Dropbox, and Google Drive provide easy exfiltration channels that malicious actors can use to upload data, especially as many organizations permit the accessing of those websites.

Steganography can be utilized in this manner by hiding large amounts of data within images and then exfiltrating those images rather than the data itself. This conceals the true nature of the exfiltrated data and makes it easier for a malicious actor to justify if confronted [13].

Ransomware While ransomware has typically focused on encrypting a target's system and demanding a ransom to decrypt all data on the infected system, data exfiltration is now being used to threaten victims into paying the ransom. Malicious actors are not only encrypting a target system's data, but they are also copying large amounts of data from any systems beforehand and publicly releasing it online if they do not receive payment.

Regardless of whether a malicious actor wishes to demand payment or not, this provides an effective method of exfiltrating data remotely, an example of this is Maze, which is a recent ransomware that was used to exfiltrate and encrypt data on any systems that it was installed on [14].

Phishing Phishing is another attack method that can be used for remote data exfiltration, performed by a malicious actor that sends several e-mails to different organization addresses to gain access to sensitive information. The malicious actor is not required to install anything on a target system but can utilize social engineering techniques such as masquerading as a trusted entity to coerce a target into divulging sensitive information such as login credentials or personal information.

By spoofing the e-mail address of a trusted entity, a malicious actor can easily be deemed trustworthy, and it means that targets will fall under the assumption that they are receiving legitimate communication from a trusted source when in reality they are communicating with a malicious actor. Spear phishing is an example of phishing that is directed at specific individuals or companies, with malicious actors gathering details and personal information on their targets to increase the probability of success [15].

Encapsulation Encapsulating data is another method that can be used to exfiltrate data remotely without detection. This is done by a malicious actors first encapsulating the targeted data into a specific file type, and then proceeding to exfiltrate the data from an organization's internal network to an external destination. File types such as ZIP, RAR, and CAB can all be used to encapsulating targeted data into an archive, protecting it from any traffic monitoring services or data loss prevention strategies that an organization might have in place. Nested zip files can also be used as many data loss prevention (DLP) systems will stop scanning nested zips after 10 to 100 archives to avoid zip bomb attacks [16].

Protocol Exploitation Many companies utilize a vast number of services and protocols to establish a working network infrastructure. So by examining the protocols being used by a company's network and identifying any vulnerabilities, a malicious actor can exfiltrate data remotely. An example is domain name system (DNS) tunneling, which can be used for data exfiltration. By establishing a C&C channel over DNS, a malicious actor can encode data from other programs and protocols into queries and responses. However, this requires a malicious actor having access to the internal DNS server as well as having an authoritative domain server setup to establish the tunneling from the internal compromised system, through the internal DNS server and to the malicious domain server. Without this, any traffic from the compromised system that is sent directly to the malicious authoritative server can potentially be blocked by the network's firewall [17].

3.2.3 Insider Data Exfiltration Threats

Insider threats refer to any malicious threats that come from the personnel who work within a company, such as employees, contractors, or former employees. These types of threats are difficult to defend against as insiders within a company are the same people who are trusted with handling confidential or otherwise sensitive data and knowledge of the company's network infrastructure. This means that all forms of insider threats must be identified, and the relevant strategies clearly defined on how to prevent insider attacks from occurring, as anyone who has access to a company's network can be considered an insider threat.

3.2.3.1 Types of Insider Threats

Compromised Insiders If an insider within a company has their credentials compromised, this can lead to a security breach caused by anyone who uses those credentials to gain unauthorized access to any confidential information that the insider might have clearance to access. However, if they do not realize that they have been compromised, then this leads to malicious actors gaining continued access to restricted data and potentially compromising more insiders within the company.

This type of threat can occur when insiders click on malicious links sent by phishing e-mails posing as legitimate and trusted sources, which result in an insider inadvertently compromising themselves and their system to an attacker [18].

Negligent Insiders Negligent insiders refer to any insiders that do follow security best practices and/or any practices which are enforced by a company to mitigate data breaches and compromises, leaving themselves and the company at risk.

Examples of this type of negligence include leaving a system unlocked while it is unattended, keeping passwords written down in open view, and using default credentials for services and software, which opens up the opportunity for an attacker to access a system and conduct an attack.

These types of insiders can easily become prime targets for attackers, especially if their negligence is spotted and monitored by other insiders with malicious intent, with 63% of reported incidents in 2020 so far caused by insider negligence [19]. Attackers can utilize attacks such as social engineering or phishing to coerce these types of insiders into divulging key information, which can prove especially effective if the target is neglectful of security issues and practices.

Malicious Insiders Malicious insiders are those people within a company who intentionally wish to cause damage or disruption to a company's network or infrastructure. These can be insiders who take advantage of their authorization or clearance to access otherwise restricted data or systems and then proceed to perform an attack, such as the exfiltration and public distribution of confidential data. Also, as these insiders are legitimate users and will understand how a company's systems are structured, they will be able to more easily identify vulnerabilities which can be exploited. This advantage leads to them being able to more easily cover their tracks and hide any clear evidence which might lead to the attack being detected.

Malicious insiders also have the capability of more severely impacting systems as they can identify key systems and data to target for an attack. Even an attack as simple as the deletion of data, systems becoming unusable, or a company having to suspend normal operation, all result in large costs due to damages and the downtime caused by them [20].

Commonly Targeted Data The type of organization and resources available is very important, as this can be a defining factor in what data might be available for an attacker to target and the possibilities involved with what can happen postattack. This is why many attackers will evaluate and determine which types of organizations are more valuable targets, taking factors into accounts such as scale, purpose, and infrastructure to pick a prime target for attack.

An example of this type of targeting can refer to companies that offer financial services such as banks and other financial establishments. This is because they are often targeted as they are responsible for the safeguarding and management of thousands of records which contain sensitive information such as cards and accounts details.

Other forms of data such as personal health information and intellectual property make prime candidates for attackers, as these can be used for blackmail,

Table 3.1 Examples of large-scale data breaches

Victim	Year	Sector	Impact	Data exposed
Yahoo	2013	E-mail	3 billion accounts	Name, DOB, contact information, and passwords
Aadhaar	2018	Government	1.1 billion accounts	Name, account numbers, photographs, thumbprints, retinal scans, and bank details
First American	2019	Financial	885 million records	Bank account, social security no. and mortgage records
Marriott International	2018	Hospitality	500 million records	Guest names, contact and passport data, and payment card numbers and expiration dates
Verifications.io	2019	E-mail	763 million records	Names, contact information, address, social media accounts and financial information

released to the public, or sold to willing buyers, especially when concerning intellectual property. The primary source of concern involving the theft of intellectual property is the direct competitors of an organization, with it attributing to 23% of the overall concern [11]. This can lead to attacks involving organized crime being more prevalent as competitors can employ a variety of tactics to acquire confidential information which may enable them to become a leader in their respective field through the use of the stolen property.

Examples of recent data breaches are shown below in Table 3.1, highlighting the impact of the breaches as well as the variety of organizations targeted and the data exfiltrated from the attacks.

3.2.4 Approaches to Counter Data Exfiltration

3.2.4.1 Preventative

Preventive countermeasures to data exfiltration involve placing measures or systems in place which attempt to block any methods that could be utilized to begin an unauthorized outgoing data transfer out of a network [21]. This can be linked to organizations that employ DLP strategies, which involve the implementation of several different systems to monitor and block a variety of techniques which could cause the loss or leakage of any confidential data [22]. Systems such as IDS and firewalls are very commonly used in this manner to monitor and verify any outgoing or incoming traffic. So if an external malicious actor attempts to access the internal network, as they are connecting from an unauthorized device/IP address, they will be blocked from accessing any of the networks resources or systems, regardless of whether they have been able to gain a physical connection to the network.

However, while both IDS and firewalls can monitor and protect an entire network, antivirus software is used for individual system protection, which can help to identify any infected systems which might have become compromised or have been targeted by external threats attempting to gain access through attacks such as phishing or masquerade attack.

3.2.4.2 Detective

Detective countermeasures focus on detecting exfiltration attempts but do not proactively prevent them from occurring [23]. While preventive countermeasures are more proactive in their defense of the exfiltration from occurring, detective countermeasures only react when an unauthorized exfiltration attempt is made. This can be focused on two levels, the first being the network level. When implemented at this level, network traffic is monitored, and some form of packet inspection is incorporated into the process so that any attempt of exfiltrating sensitive information, whether it be masked or not, can be inspected and spotted before it can leave the network [24].

The second level that this can be implemented into is the host-level, which can focus on monitoring the user access patterns while they are using a system or any data transfers that have been initiated. Through the monitoring of system access and any abnormalities in the way a system is used, such as a user attempting to access multiple file shares they are not permitted to access or a user attempting to copy a large number of files to an external storage device, alerts can then be triggered and sent to the relevant personnel, which can aid greatly in the identification of any malicious insiders [25].

3.2.4.3 Investigative

Investigative refers to countermeasures which are deployed to investigate data exfiltration incidents after they have occurred, with the goal being to investigate and determine key information from the attack to prevent it from happening again. By identifying how, when, and who exfiltrated data, it can not only aid in the identification of the person responsible for the attack but also what methods were used, or vulnerabilities exploited to be able to perform the attack. Investigating incidents can also aid in the mitigation of any effects in the aftermath of an attack which might potentially harm a company or its operation. Any details found can be key in leading investigators to determine the potential impacts that can occur and from there, take the necessary steps to mitigate them or minimize the impact they will have.

Investigative countermeasures can also be useful for other companies and organizations in a large range of different sectors, as victims of a recent attack can disclose key information. Areas such as how the attack was performed, what vulnerabilities were exploited, and most importantly, the recommended security patches or fixes that should be made to help prevent an attack of that nature from occurring again. If a victim of a recent exfiltration attack decides to disclose information on the attack, all confidential security or personnel information must be omitted to ensure confidentiality; otherwise, this can result in further data being leaked [25].

3.2.5 Mechanisms to Defend Against Physical Data Exfiltration

3.2.5.1 Network-based
Firewall A firewall is one of the most common methods used to prevent data exfiltration, as it can be utilized to monitor incoming and outgoing communications as well as allowing or denying certain communications depending on port, IP address, and file type. This means that in a scenario where a malicious insider has physical access to a system and attempts to exfiltrate data from the internal network to a cloud storage service [26–28], a firewall will inspect the traffic before it leaves the internal network and from that, determine that it is outgoing to an IP address or service that is not permitted and will be denied.

User Authentication and Privileges Authentication is a network-level mechanism that can be used to counter data exfiltration [29]. This places a barrier between the attacker and a target system as they must acquire valid credentials to authenticate with the system, regardless of whether the attacker has physical access or not.

This works well even if the attacker is a malicious insider who has their credentials, as it does not guarantee that they will be able to access all the data on the system if they do not have sufficient permissions or privileges to access anything other than their data. Systems such as Active Directory allow for this to be done, with privileges having the ability to be set for entire groups of users or on a per-user basis, meaning authentication and privileges can be fine-tuned so only authorized personnel can access certain programs, files, and perform specific actions.

3.2.5.2 Physical-based
Locks Locks are very important in helping to prevent physical exfiltration as they prevent any malicious insiders from being able to steal devices and exfiltrate data away from the premises. Examples include Kensington locks for laptops and computers, which can provide a simple prevention mechanism that stops any unauthorized personnel from moving the systems or stealing them. While some locks might be able to be broken, it is less likely that a malicious attacker

would break a lock as this would result in the incident being promptly detected, potentially leading to the attacker being identified. Apart from being locks for the devices themselves, door locks can prevent unauthorized personnel from entering restricted areas such as server rooms, which if accessed, can allow an attacker to not only disrupt a network's operation but also install malware or modify certain network properties to aid them in performing future attacks.

Physical Authentication and Biometrics   Physical authentication is another method that can be used for defense against physical data exfiltration attacks, using authentication methods such as fingerprint, facial recognition, and smart card reading that can all help to provide physical authentication to ensure only permitted users can access their systems. This can also be used to support already-established authentication on systems in the form of two-factor authentication. By inputting a username and a password, a user can then be prompted to insert an ID smartcard or scan their fingerprint to provide another layer of authentication. This type of authentication can make it nearly impossible for malicious actors to be able to access a system without the correct credentials and biometrics or identification.

Companies or organizations that deploy enterprise-level devices to its employees for enabling working during travel and at home benefit highly from this form of authentication as it ensures that even if a device is stolen or lost, then it can be assured that no one will be able to access the system or any data on it.

3.3   A Defence Mechanism for Physical Data Exfiltration Mitigation

Figure 3.1 shows the proposed physical data exfiltration mitigation mechanism. The main design feature of this solution is that the management server is used to not only store but also create both data fingerprints and the rules used for any endpoints in the network. Only network administrators would be permitted to access the server, with only one set of valid credentials being used to access the server. This aids in minimizing the number of credentials that could become a target by a malicious insider, who might attempt to steal or acquire the credentials to gain unauthorized access to the server. Algorithm 1 summarises the operation of the proposed physical data exfiltration mitigation mechanism.

3.3.1   Confidential Data Identification

Confidential data identification involves the use of different techniques to automatically determine whether files contain any confidential information. When

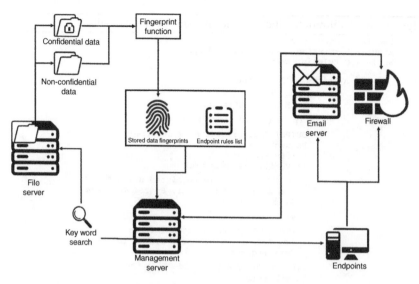

Figure 3.1 A component diagram of the proposed physical data exfiltration mitigation mechanism.

initiated, each file hosted on a system or a network will be examined to identify if it is considered confidential or nonconfidential, scouring all directories and storage to identify as many items as possible. RegEx and keyword searching are some of the most common techniques used to identify sensitive data, with RegEx focusing on expressions and the context of the data while keyword searching identifies specific words that have been identified as representing confidential information.

This described mechanism utilizes keyword searching for the identification of any confidential information, with the management server being used to scan through all network-hosted data to identify and place each file into a category of either "confidential" or "nonconfidential." The server can be configured with a keyword search list in which administrators can add to or modify any preexisting lists that are already contained on the system. This makes the keyword search much more versatile, as it is capable of being altered to meet the needs of a variety of different organizations, regardless of their function or the type of data that they handle.

Figure 3.2 illustrates that once the server has identified whether a file is confidential or nonconfidential, the result is stored until the corresponding file fingerprint has been created. Once the fingerprint has been created, the corresponding result is placed on the fingerprint record. This means that when performing a confidentiality lookup, the server can examine the relevant fingerprint record and quickly identify whether it has been tagged as confidential or not.

Algorithm 1 Exfiltration mitigation mechanism operation details.

```
Require: Action(A), File (F), Fingerprint(Fp)
Ensure: Permission (Deny or Allow Request)
While A ∈ Rulelist then

    If A status = 'Allow'then

        Fp = SHA256(F)
        If Fp ∈ FingerprintList then

            If Fp is confidential then

                OUTPUT "Authorisation is required to use"
                Request = F + A
                Verify Request
                If Verify  returns true

                    OUTPUT "Action is permitted using " + F
                    Log Action, File, User, System ID, Date
                    Permit Request

                Else

                    OUTPUT "Action is not permitted using " + F
                    Log Action, File, User, System ID, Date
                    Deny Request

                Else if Fp is not confidential then
                Permit Request
                Log Action, File, User, System ID, Date

            End if

        Else if Fp ∉ FingerprintList then

            KeyWordSearch(F)
            OUTPUT "File does not match any on system, try again
            latter"
            Deny Request

        Else if Fp not present then

            Log Action, File, User, System ID, Date
            Permit Request

        End if

    Else if  A status = "Deny" then

        OUTPUT "Action is not permitted"
        Log Action, File, User, System ID, Date
        Deny Request

    End if

End While
OUTPUT "Action is not permitted"
Log Action, Unknown Tag*, File, User, System ID, Date
Send Log
Deny Request
```

3.3.2 Endpoint Access Rules

As many threats posed by malicious insiders reside from physical threats involving the use of internal systems to perform malicious actions, there must be a level of defense on a per-endpoint basis to ensure that physical threats can be dealt with while the mechanism resides on the network-level. Endpoint access rules allow for that level of defense, by deploying a rule list on a network, and it ensures that

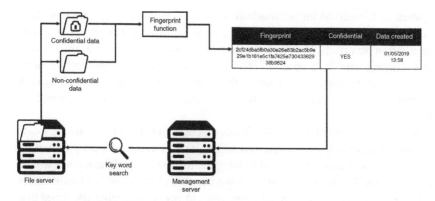

Figure 3.2 Confidential data identification flow.

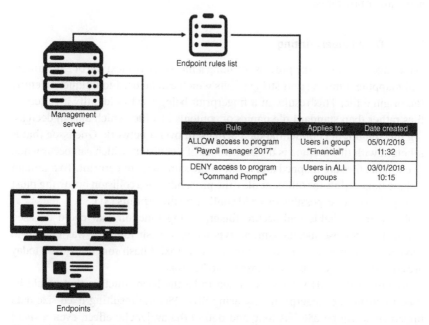

Figure 3.3 An illustration of endpoint rule list to mitigate unauthorized activities.

all systems on the network must follow a set number of rules to mitigate the unauthorized activity (see Figure 3.3).

Utilizing the least-privileged required zero-trust principle ensures that the
minimum amount of permissions is granted to users to ensure that they can
use their systems to perform any required tasks while ensuring that any other
actions are prohibited [30]. This can include prohibiting the use of in-built

system tools reserved for administrators such as system settings, a terminal, or command prompt, which is accessed by a malicious insider that can result in data exfiltration occurring. Also, it can be used to prohibit actions such as transferring and copying data from specific directories and preventing unauthorized users from accessing programs outside of their respective department, such as only permitting access to payroll software to those users who work within the financial department.

If an action request is sent to the management server, but it is not able to locate a corresponding rule for the received action, it will immediately deny the action from being carried out. As well as denying the request, a log of the request will be created but will contain a special unknown tag which is then sent to administrators for review. This is done to alert administrators to any actions being attempted that have not yet been accounted for, such as the attempted running of newly installed software or programs.

3.3.3 Data Fingerprinting

Data fingerprinting is the process of taking a file, such as a text file or a document and mapping it into a short string of bits which can be used to uniquely identify the original file. This results in a fingerprint being used to identify a variety of files rather than standard file names or contents of a file, which aids in decreasing the amount of confidential data being sent over a network. One issue that is inherent with this method is the occurrence of a collision, which can occur when two unique files are mapped and result in the same data fingerprint. To maintain unique fingerprints for different files, the probability of a collision occurring must be kept to as low as possible to avoid conflicting fingerprints [31].

To create a reliable and secure fingerprinting function that could be used within this defense mechanism, a cryptographic hash function was chosen to serve as a fingerprint function, as many mainstream hash functions used today are considered very secure and have yet to be broken.

Because of this, SHA-256 was chosen to be the hash function that would be used to create the fingerprints. By using SHA-256, the resulting fixed-size data fingerprints will be 256 bits long and due to the avalanche effect, even a small change in the original file will result in a new hash, helping to avoid any potential fingerprint collisions. Also, as it is a one-way function, it becomes much more challenging for any attacker, whether they are an external attacker or a malicious insider, to decipher what information is being referenced from the hash alone.

This is a slower method compared to other fingerprinting techniques such as Rabin's fingerprinting algorithm, which is designed specifically for the task of mapping data fingerprints and analyzing collision probability. However, Rabin's algorithm is not secure against malicious attackers as the use of an internal key during fingerprint generation means that if an attacker can locate and acquire the key, they gain the ability to modify files without causing changes to the fingerprints [32].

In this mechanism, the fingerprints are created and then stored on the management server. This is done so that when a user on an endpoint requests to copy, modify, or send a file, a request will be sent to the management server asking to verify whether the file involved in the request is classed as being confidential or not. By examining the request and cross-referencing the hash function to the file that has been requested by the user, it can then determine whether the file is confidential or not and either permit or deny the user the ability to perform that action.

3.3.4 Relevance to Physical-Layer Protection

While this mechanism is a network-based one, it still has relevance to physical-layer protection as it is capable of safeguarding from physical-layer threats such as the unauthorized access or use of a system and confidential data. By utilizing an endpoint rule list, it can help to create a universal defense against malicious or unauthorized use of any systems within the internal network, with each system following a set of rules with only permitted actions being performed. An example would be if a malicious insider attempts to transfer data from the network to an external storage device, with the action being denied by the management server, as it is not on the list of permitted actions.

Physical-layer protection is also reinforced using network-based confidential data identification, with all network files and data being put against a keyword search to determine the level of sensitivity which is within each file. This means that even if an action is permitted by the management server, such as sending an e-mail, if the corresponding file/files used are found to be tagged as confidential, then the action will be denied. This helps to prevent malicious insiders from using otherwise valid actions to exfiltrate confidential data, as both the action and file are verified.

3.3.5 Complementing Existing Firewall and Application-based Measures

To work harmoniously within a variety of different network infrastructures, the mechanism needs to be able to work seamlessly with any already-existing deployments that are used to aid in the prevention of data exfiltration. Typically, both firewalls and application-based methods are used to help prevent data exfiltration attacks from occurring, with firewalls focused on monitoring and verifying all communications and application-based measures being utilized to help prevent personnel from performing any prohibited actions within any applications.

As application-based defense measures are bound to a single application, they can help to protect applications from being used in malicious or prohibited ways along with ensuring that only personnel with the proper privileges can perform certain tasks. However, this can still leave an endpoint at the risk of being used in a malicious manner outside of any application. Endpoint rules complement application-based measures as they ensure that there are restrictions and defenses in place to stop not only prohibited actions within applications but also on the system as a whole. By deploying a rule list on all network devices, it ensures that users are given the least privilege required to mitigate any attempts of maliciously using a system without hindering a user's ability to complete tasks efficiently. Figure 3.4 shows the endpoint rules and application-based measures.

As a typical firewall focuses on examining traffic as it leaves and enters the network, they tend to have their own established set of rules to permit and deny a variety of different traffic types, ports, and addresses. This can work well with a confidential data identification method, as the confidential or nonconfidential tag can be used to initially verify whether data being used in an action contain any sensitive information. Then, once it has been verified to contain no sensitive information or there has been permission granted for the sensitive information to be used, the firewall must verify that the traffic is permitted to leave the network by examining properties such as destination and source address, port, and application. This results in an extra layer of security, as even if both the

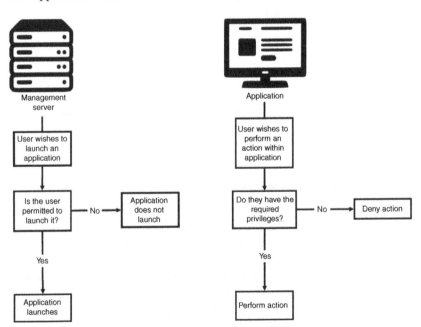

Figure 3.4 Endpoint rules and application-based measures.

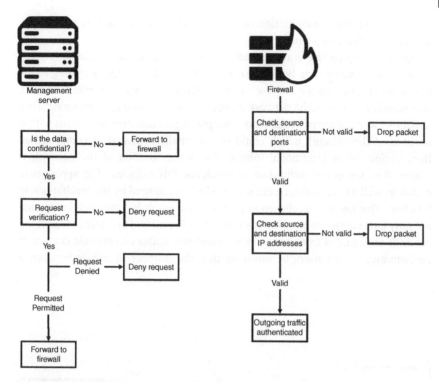

Figure 3.5 Work flow for confidential data list and firewall deployment.

endpoint rule list permits the action and confidential identification allows the corresponding data to be used, if the firewall identifies that the traffic is not permitted to leave the network after examining it, then it will be denied. Figure 3.5 shows the confidential data list and firewall deployment.

3.4 Implementation and Analysis

3.4.1 Experimental Setup

To test the proposed mechanism and its implementation into an already established network, a simulation tool will be used to create a simulated network containing a variety of systems and components that are found within a business networks' infrastructure. A simulated network will be used as this helps in more accurately depicting the process of implementation as well as demonstrating the defense capabilities of the mechanism once it has been implemented. The Graphical Network Simulator [33] was chosen to create the simulated network infrastructure as it has a large variety of tools and options that can be used to create an accurate representation of a real-world network that this mechanism would be implemented within.

Figure 3.6 depicts the simulated network layout that features several different areas which have all been both physically and logically isolated according to the business operation sector and use of the systems. This includes HR, Financial, and Admin portions of the network all being kept separated to improve network scalability and reduce potential risks from placing administrative systems within portions of the network which could be potentially breached or compromised from accidental or intentional misuse. The server portion of the network is where all networks key systems are situated, and this includes a file, application, e-mail as well as the management server which is required by the mechanism to function. This layout was chosen as it provides a simple but efficient design that not only has security in mind but also ensures different and unrelated areas of the network are isolated from one another. However, it also can provide consistent performance for all users by ensuring that the number of points between a

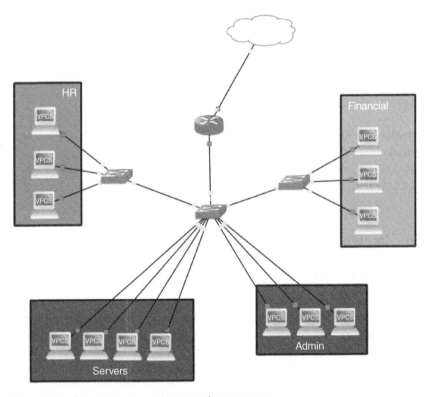

Figure 3.6 Simulated network layout and components.

users' endpoint and the key systems is kept to a minimum to reduce potential bottlenecks and reductions in bandwidth.

For how the mechanism would be integrated into the network, the management server would be situated within the server section of the network, pairing it alongside all other servers within the network that provide key functions for the network and its operation. This ensures that all key systems are situated together in one "zone" and are physically isolated from other areas of the network to help in the mitigation of any attempts of malicious use or disruption to the network.

3.4.2 Threat Scenario

The scenario that will be used in conjunction with the simulated network focuses around a malicious insider within the financial department of the company who has decided to attempt to exfiltrate some confidential data to sell it on to the competitors of the company. As the malicious insider works within the financial department, this means that he or she immediately has access to a lot of sensitive information regarding all the spending and projected profits for the company, which is prime target for data exfiltration attacks. To try and minimize the risk and chance of being spotted by other employees and administrators, the insider has chosen to exfiltrate only one file by transferring it to a USB storage device that the insider has been able to conceal and get onto the premises. The company that the malicious insider is employed at has a strict policy on the use of removable storage devices and forbids the use of them for all employees.

Using this scenario will allow for the approach taken by the mechanism to be shown in detail on a step-by-step basis, showing how it is capable of mitigating physical data exfiltration attacks while being a network-based mechanism. Figure 3.7 shows the threat scenario.

3.4.3 Scenario Execution and Analysis

To begin, the malicious insider will first access the file server on the network and identify the file that he/she wishes to exfiltrate from the network to the removable storage. As he/she works in the financial department, he/she has been able to identify a spreadsheet that contains a large amount of information on the company's current spending and projected earnings for the financial year.

Once a suitable target has been chosen, the malicious insider will then attempt to begin the process of exfiltration by inserting their removable storage device into a USB port on their system. Once this is done, the mechanism will immediately take over and begin by first identifying the action that has just been taken, which in this instance is that a device has been inserted into the USB port. Once the mechanism has identified the action taken, it will immediately halt the corresponding process

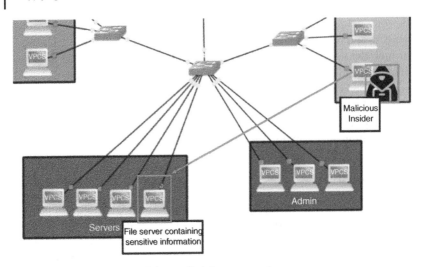

Figure 3.7 An illustration of the studied threat scenario.

(e.g. necessary drivers being loaded to use the device) and will form an action request to be sent to the management server.

When creating the action request, the mechanism will use several key information including the IP address, user account, and the workstation that the action request originated to maintain a consistent level of nonrepudiation no matter where or when an action is requested. As seen in Figure 3.8, an action request

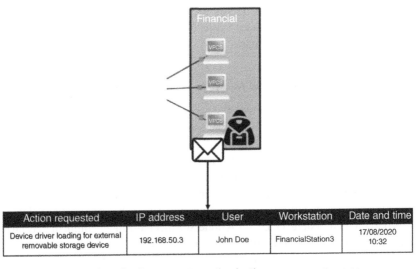

Action requested	IP address	User	Workstation	Date and time
Device driver loading for external removable storage device	192.168.50.3	John Doe	FinancialStation3	17/08/2020 10:32

Figure 3.8 Illustration of action request creation by the management server.

has been created using those key items as well as defining the specific action requested which is the loading of a device driver for an external storage device. An important note to make is that if the malicious insider was to perform an action that involved the use of a specific file, then an extra item would be attached to the request which would refer to the file or files used. This is so that the management server can also check whether the file or files used contain any confidential data, even if the action is permitted.

When the action request has been successfully created, it is immediately sent to the management server, which will inspect the action request and identify what the action is contained within the request. Once it has identified the action, it will then attempt to find the corresponding action within its endpoint rule list, which is used to determine what actions can and cannot be performed by users on their systems.

When a corresponding rule has been located, the management server will identify whether the action has been permitted or denied, and depending on the result, it will send a response back to the relevant endpoint informing it of whether to resume or cancel the action. In this scenario, as the company has a strict policy on the use of external storage devices, the management server will find the corresponding rule and identify that it is an action that is denied.

After the action has been identified as being denied, an action response is created which includes the action requested, the result (permitted or denied) as well as a date and time stamp, as seen in Figure 3.9. This is logged by the management server and then sent back to the malicious insiders' endpoint, where it will inspect the response and determine whether it is permitted or denied permission to continue the requested action.

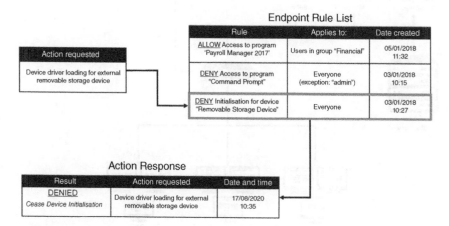

Figure 3.9 Endpoint rule search and action response creation.

As there is a denied marker within the response, the endpoint will immediately cease the initialization of the device, leaving the external storage device unusable. This results in the physical data exfiltration attack being prevented, as the malicious insider is no longer able to utilize the external storage device to transfer confidential data from the internal network.

3.5 Evaluation

To evaluate the effectiveness of the proposed mechanism, we place it in several different hypothetical scenarios. Each scenario will focus on a specific physical attack method to showcase if, and how, the mechanism can mitigate a variety of physical data exfiltration attacks. Using a variety of physical attack methods for this evaluation is crucial as it helps to determine the ability of not just the mechanism as a whole but also the individual components that are a part of the mechanism.

To maintain a fair and unbiased evaluation method, several assumptions will be established beforehand, with the assumptions being created to reflect the general functionality of a typical business network and how each key system functions. Also, to complement these assumptions, each scenario will be based within an assumed network environment, that will be used to represent a typical business network infrastructure seen in Figure 3.10.

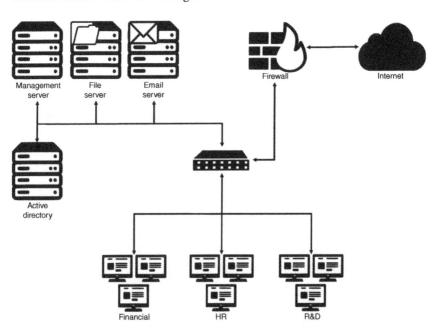

Figure 3.10 Assumed scenario network topology and studied network components.

The assumptions are as follows:

1. Employees have physical access to their systems.
2. All data is hosted on a file server.
3. Employees are prohibited from using external storage devices.
4. Users' file and program access privileges are managed separately via Active Directory
5. Users in one department are not able to access files belonging to another department.

3.5.1 Scenarios

Table 3.2 shows the three scenarios used to evaluate the proposed mechanism and their ability to mitigate physical data exfiltration attacks. These scenarios were chosen as they all feature an attack method which tests a particular component of the mechanism, including the confidential data identification and endpoint rule list. Each scenario uses a different attack method; however, all of them are based on the idea that a malicious insider will be performing the attacks within the assumed network, intending to exfiltrate sensitive information.

To ensure that the scenarios used for the evaluation were relevant and could properly test the effectiveness of the mechanism, the objectives of each scenario directly relate to a specific design objective that was established earlier in the design process. By creating each scenario and correlating it to a design objective, it ensures that each scenario properly evaluates a key area of the mechanism; so an accurate determination can be made of how the mechanism can perform and how it can meet the series of objectives that were set.

Table 3.2 A summary of the scenarios used in the proposed mechanize evaluation

Scenario	Description	Objective
1	Malicious insider attempts to use an external storage device to exfiltrate confidential information	Identify how an endpoint rule list can be used to block the unauthorized transfer of data to an external storage device
2	A malicious insider with valid credentials/access to confidential data attempts to upload to cloud storage	Confirm that confidential data identification can be used to stop exfiltration attempts to remote networks, even if the insider has valid access to the confidential data
3	Malicious insider attempts to use a malicious python tool to create a backdoor into the network for data exfiltration	Identify if the mechanism is capable of mitigating the use of malicious programs/tools

As well as providing an in-depth description of how the mechanism mitigates the corresponding attack method in each scenario, a flowchart will also be used to provide a bird's-eye view of the step-by-step process taken to mitigate the data exfiltration attack.

3.5.2 Scenario 1

The first scenario focuses on a malicious insider who is attempting to utilize an external storage device to physical exfiltrate data from the internal network. As many external storage devices such as USB are very small in size, they are easy to not only transport but also conceal from other people. This scenario will begin with the malicious insider plugging in the external storage device into an available USB port on their system. Once this has been done, the mechanism will immediately takeover thanks to the endpoint rule list that has been deployed on the network.

The system will first halt the initialization of the storage device and send an action request to the management server to verify whether initialization of a new USB device is permitted. The management server will then check its endpoint rule list to determine whether there is a corresponding record for such an action, if there is, then it will check to see whether it is to be permitted or denied. Because the use of external storage devices is prohibited on any systems within the internal network, the management server will identify the action as being denied and will send a response back to the endpoint alerting it that it cannot continue that action and will cancel the initialization of the device, leaving it unusable. A log will also be created and stored on the management server. Which will contain key information on the recent request, including the action that has been denied, name of the user who sent the request, and endpoint it was sent from along with the time and date.

As this is a mechanism that follows the zero-trust principles, any action that does not have a corresponding record on the management will be immediately denied [6]. This is to ensure that malicious insiders are not capable of using any programs, features, or other items maliciously before they are integrated into the rule list.

3.5.3 Scenario 2

The second scenario focuses on a malicious insider attempting to use a cloud storage service to exfiltrate data by directly uploading the target data from the internal network. As cloud storage services such as Dropbox, Google Drive, and OneDrive have become increasingly more popular for the storage of information, it makes them an ideal tool to use to exfiltrate data [34]. However, while they are used widely by standard consumers, depending on the security policies that are in place within a business, they can either be restricted or outright prohibited.

To begin, the malicious insider will first identify a cloud storage provider to exfiltrate the data to, as well as identifying what data will be targeted for exfiltration. Once the insider has been able to access a cloud storage website and has chosen a piece of confidential information to exfiltrate, he or she can begin the process of exfiltration by attempting to upload the file. Before the file can be uploaded, the mechanism will initiate by sending an action request to the management server. As the action being performed involves the use of a file, before it is sent to the management server, it will hash the file using SHA-256 to create a file fingerprint.

Once the fingerprint has been created, the request is sent. It will first query the management servers' endpoint rule list to determine if the copying of a file is permitted, which it is. However, as the request also includes a file fingerprint, the management server will take the file fingerprint and try to find a corresponding fingerprint within its list of stored fingerprints to determine whether it is confidential or nonconfidential. The management server will locate the corresponding fingerprint record and by checking the confidential tag assigned to the record, it will be able to determine that it is a confidential file and therefore cannot be used without verification. It will then prompt the insider that verification is required to perform this action using this file and will give them the option to either ask for verification or cancel the action.

Verification is used in this stage as in the event where confidential data have to be used legitimately such as being sent in an e-mail to a trusted contact, it can be verified and then sent. A verification request is similar to an action request; however, it is sent to an administrator for review so they can determine whether it is a legitimate use of confidential data. In this scenario, even if the insider chooses to send a verification request in the hopes that it will be permitted to proceed, as the file used is confidential and the use of cloud storage is prohibited, it will be denied. If the file was not confidential, then it would be allowed to proceed to the firewall as which would be responsible for verifying other information such as the source and destination IP addresses and ports along with a packet inspection.

3.5.4 Scenario 3

This scenario focuses on a malicious insider who has been able to acquire a malicious tool that can be used to set up a backdoor into the internal network so crucial confidential data can be accessed remotely. The insider has stored the malicious tool within their downloads folder; however, the malicious insider must run the tool to begin the process of installing the backdoor.

Once the insider attempts to run the tool, the mechanism will initialize and begin by halting the file from being run and creating an action request. The action request will be sent to the management server, where it will be used to determine whether the running of a python file is permitted. As this is a python-based tool which uses the ".py" extension, the management server can check its endpoint rule list for a rule relating to files with that particular file extension.

If the management server can locate a rule with the corresponding file extension, it will then determine whether to permit or deny the action of running the file. However, as per the established assumptions mentioned earlier, there is currently no rule on the management server which dictates the use of python files. This results in the management server not being able to detect an associated rule and will therefore immediately deny the request.

The management server will also create a log for this request; however, it will contain an unknown tag on the request as it was not able to find an associated rule for the request and was therefore required to immediately deny. Once an administrator receives the unknown request log, he or she will be able to review it and determine whether a new rule should be added to the endpoint rule list and whether the corresponding action should be denied or permitted. This process of reviewing unknown requests is very beneficial as it helps administrators to not only spot malicious activity and create new rules to prevent it but also to identify issues where users are not able to perform legitimate tasks such as running a program due to a missing or incorrectly configured rule.

3.5.5 Results Analysis and Discussion

Examining the results from the evaluation shows that the mechanism is capable of mitigating several different variations of physical data exfiltration attacks, while being completely situated in the network layer. Physical-based defense methods, such as locks, covers, and enclosures, can be used to help mitigate physical data exfiltration attacks by restricting access to the systems; however, they cannot account for any malicious use that occurs on the system itself and is prone to falling through tampering. This is why a network-based defense method offers the best of both worlds and can be used to not only stop malicious network activity but also can be utilized to stop malicious physical activity from succeeding.

By having a dedicated management server to host all of the core components of the mechanism along with the endpoint rule lists and file fingerprints, it restricts the number of ways a malicious actor could potentially access and disrupt the operation of the mechanism. This helps to reinforce zero-trust principle, as it assumes that anyone could be a threat, so by self-containing all the components and data onto a single point, it offers a secure method of operation without hindering performance. This does result in a single point of failure, but this is a small price to pay in comparison to the large number of potential breaches that could occur if the mechanism was spread out over the network.

The versatility of the mechanism is another advantage as it is capable of being configured in many ways to fit the environment that is being deployed in. Endpoint rules can be configured in any way an administrator sees fit and can be used to segregate access of programs, settings, and other actions depending on the different user groups on the network, so users are only able to access what they need, be able to perform their required tasks, and nothing more. Keyword scanning can also be configured so that the determination of whether files contain confidential data or not can be changed based can what keywords are indicated as confidential. This is crucial as depending on the type of business the mechanism is being deployed in, confidential information can vary and with that so can the keywords associated with what is confidential and nonconfidential.

Logging is a very important factor in helping to not only troubleshoot issues but also to help identify any indications of attempted malicious activity. The logging of all permitted and denied requests is massively beneficial as it helps administrators to monitor user activity and to verify that both the endpoints and their users are following the rules and policies that have been set. Also, the use of unknown requests helps administrators to identify any potential loopholes or areas which are being exploited, with all actions that do not have a rule associated with them being explicitly denied until a rule has been created. This is especially useful as there are occasions where vulnerabilities are not properly accounted for and can be potentially exploited by attackers.

One of the main limitations is the actual feasibility of it being deployed onto a network. As this is a proposal with no developed prototype to evaluate, no formal testing has been performed on any real hardware, which means the feasibility of the mechanism can only be estimated and not properly measured. While there can be an evaluation of the mechanism in a general, hypothetical sense, this does not consider any of the network-related issues, implementation problems, and the general performance that could be seen when implemented on real hardware.

Another limitation is that it has a single point of failure. By having all components and data housed within a single point, if it is breached, the attacker will have access to the entire mechanism and all the information on its operation, including the stored endpoint rules and file fingerprints. This can result in the attacker manipulating the mechanism to their favor, potentially resulting in large amounts of data being exfiltrated and damage to the business. Also, if the management server goes offline, the entire network must be taken offline to help mitigate any attempts to exfiltrate confidential data while the server is offline. This can potentially result in considerable amounts of downtime until the management server is operational again, causing considerable damage due to network operation being halted.

An improvement that could be made to this mechanism is the merging of firewall packet inspection and file fingerprints stored on the management server.

By using the file fingerprints, not only could a firewall inspect a packet and the source and destination addresses, but it could also verify the data attached to a packet does not contain any confidential data. This would remove the need for the management server to perform the file fingerprint check, helping to reduce the number of verifications being performed, as the packet inspection and fingerprint check would be merged into one. This could potentially help to prevent malicious insiders from concealing confidential data within permitted traffic types, as the data would still be verified even if the protocol associated with the packet is permitted.

3.6 Conclusion

As attackers seek more ways to exfiltrate confidential data out of company networks, it is important that the security protecting the data evolves to effectively defend against the variety of different attack avenues that are used today. By creating a mechanism that revolves around zero-trust, it ensures that the defense is consistent across all areas, as there is a constant assumption that unless proven otherwise, every user is a potential threat and requires constant verification. This is key when looking at the types of attacks that are occurring at the hands of malicious insiders, with disgruntled employees and ex-employees using their knowledge of the inner network and its systems to find exploits to cause damage by leaking or selling sensitive information and attempting to disrupt the operation of a network.

While this work merely provides a proposal, it has shown that it is possible to develop an effective network-based defense mechanism that is capable of preventing physical data exfiltration from occurring. All while removing the vulnerabilities that have typically been associated with physical-based defense mechanisms which can be prone to tampering, ultimately leading to those mechanisms failing. In future, we plan to implement a prototype to implement and evaluate our proposal on real-hardware under realistic real-world environment.

References

1 Marques, R.S., Epiphaniou, G., Al-Khateeb, H. et al. (2020). A flow-based multi-agent data exfiltration detection architecture for ultra-low latency networks. *ACM Transactions on Internet Technology* 21 (4): 1–30.

2 Risk Based Security (2020). 2020 Q1 Data Breach Quick View Report. Data Breach Quick View Report. https://pages.riskbasedsecurity.com/en/2020-q1-data-breach-quickview-report (accessed 3 June 2020).

3 Walker-Roberts, S., Hammoudeh, M., Aldabbas, O. et al. (2020). Threats on the horizon: understanding security threats in the era of cyber-physical systems. *The Journal of Supercomputing* 76 (4): 2643–2664.

4 Mackintosh, M., Epiphaniou, G., Al-Khateeb, H. et al. (2019). Preliminaries of orthogonal layered defence using functional and assurance controls in industrial control systems. *Journal of Sensor and Actuator Networks* 8 (1): 14.

5 Do, Q., Martini, B., and Raymond Choo, K.-K. (2015). Exfiltrating data from Android devices. *Comput. Secur.* 48: 74–91. https://doi.org/10.1016/j.cose.2014.10.016.

6 Elmrabit, N., Yang, S. and Yang, L. (2015). Insider threats in information security categories and approaches. *2015 21st International Conference on Automation and Computing (ICAC)*, 1–6. https://doi.org/10.1109/IConAC.2015.7313979.

7 Bitglass (2020). BYOD: Bitglass' 2020 Personal Device Report. Bring Your Own Device Report. https://www.bitglass.com/blog/bitglass-2020-personal-device-report (accessed 6 June 2020).

8 Ghafir, I., Prenosil, V., Hammoudeh, M., Han, L., and Raza, U. (2017). Malicious SSL certificate detection: a step towards advanced persistent threat defence. In: *Proceedings of the International Conference on Future Networks and Distributed Systems (ICFNDS '17)*. Association for Computing Machinery: New York, NY, USA, Article 27. https://doi.org/10.1145/3102304.3102331.

9 Chismon, D., Ruks, M., Michelini, M., and Waters, A. (2020). Detecting and Deterring Data Exfiltration, Guide for Implementers. *Technical report*. MWR Info Security. https://www.bitglass.com/blog/bitglass-2020-personal-device-report (accessed 20 June 2021).

10 Do, Q., Martini, B., and Choo, K.-K.R. (2015). Exfiltrating data from android devices. *Computers & Security* 48: 74–91.

11 McAfee (2019). Grand Theft Data II: The Drivers and Shifting State of Data Breaches. https://www.mcafee.com/enterprise/en-us/forms/gated-form-thanks .html?docID=b5e4babd-d8f1-4155-a242-5c578ef4c6c8 (accessed 6 December 2020).

12 Ghafir, I., Prenosil, V., Hammoudeh, M. et al. (2017). Malicious SSL certificate detection: a step towards advanced persistent threat defence. *Proceedings of the International Conference on Future Networks and Distributed Systems* (ICFNDS '17). Association for Computing Machinery, New York, NY, USA, Article 27. https://doi.org/10.1145/3102304.3102331.

13 Shulmin, A. and Krylova, E. (2017). Steganography in contemporary cyber attacks. *Kaspersky Lab.*

14 McAfee (2019). Ransom ware Maze. https://www.mcafee.com/blogs/other-blogs/mcafee-labs/ransomware-maze/#_ftn1 (accessed 2 December 2020).

15 NCSC (2018). Phishing attacks: defending your organisation. https://www.ncsc .gov.uk/guidance/phishing (accessed 2 December 2020).

16 Pellegrino, G., Balzarotti, D., Winter, S., and Suri, N. (2015). In the compression Hornet's nest: a security study of data compression in network services. *SEC'15*, pp. 801–816. ISBN 9781931971232.

17 Raman, D., DeSutter, B., Coppens, B. et al. (2012). DNS tunneling for network penetration. *International Conference on Information Security and Cryptology*, 65–77.

18 Ray, T. (2014). Anatomy of the Compromised Insider. https://www.imperva .com/docs/webex/Webinar_20141119_Anatomy_of_the_Compromised_Insider .pdf (accessed 2 November 2020).

19 IBM (2020). Cost of Insider Threats: Global Report. https://www.ibm.com/ security/digital-assets/services/cost-of-insider-threats/#/ (accessed 4 November 2020).

20 Elmrabit, N., Yang, S.-H., and Yang, L. (2015). Insider threats in information security categories and approaches. *2015 21st International Conference on Automation and Computing (ICAC)*, pp. 1–6. IEEE.

21 Ghafir, I., Prenosil, V., Hammoudeh, M. et al. (2018). BotDet: A system for real time botnet command and control traffic detection. *IEEE Access* 6: 38947–38958.

22 Gibson, D. (2012). *CISSP Rapid Review*. Pearson Education.

23 Ghafir, I., Prenosil, V., Hammoudeh, M. et al. (2018). Disguised executable files in spear-phishing emails: detecting the point of entry in advanced persistent threat. *Proceedings of the 2nd International Conference on Future Networks and Distributed Systems*, pp. 1–5.

24 Ghafir, I., Prenosil, V., and Hammoudeh, M. (2016). Botnet command and control traffic detection challenges: acorrelation-based solution. *4th International Conference on Advances in Computing, Electronics and Communication*, pp. 15–16.

25 Ullah, F., Edwards, M., Ramdhany, R. et al. (2018). Data exfiltration: a review of external attack vectors and counter measures. *Journal of Network and Computer Applications* 101: 18–54.

26 Teing, Y.-Y., Homayoun, S., Dehghantanha, A. et al. (2019). *Private Cloud Storage Forensics: Sea File As a Case Study*, 73–127. Cham: Springer International Publishing. ISBN 978-3-030-10543-3. https://doi.org/ 10.1007/978-3-030-10543-3_5.

27 Shehu, U., Safdar, G., and Epiphaniou, G. (2016). Fruit fly optimization algorithm for network-aware web service composition in the cloud. *International Journal of Advanced Computer Science and Applications* 7 (2). https://doi.org/10.14569/IJACSA.2016.070201.

28 Kayes, A.S.M., Kalaria, R., Sarker, I.H. et al. (2020). A survey of context-aware access control mechanisms for cloud and fog networks: taxonomy and open research issues. *Sensors* 20 (9): 2464. https://doi.org/10.3390/s20092464.

29 Walshe, M., Epiphaniou, G., Al-Khateeb, H. et al. (2019). Non-interactive zero knowledge proofs for the authentication of IoT devices in reduced connectivity environments. *Ad Hoc Networks* 95: 101988.

30 Walker-Roberts, S., Hammoudeh, M., and Dehghantanha, A. (2018). A systematic review of the availability and efficacy of counter measures to internal threats in healthcare critical infrastructure. *IEEE Access* 6: 25167–25177.

31 Roussev, V. (2009). Hashing and data fingerprinting in digital forensics. *IEEE Security & Privacy* 7 (2): 49–55.

32 Devopedia (2020). Fingerprinting algorithms. https://devopedia.org/ fingerprinting-algorithms (accessed: 16 November 2020).

33 GNS3 (2021). Graphical network simulator. https://www.gns3.com/. (accessed 02 March 2021).

34 Belguith, S., Kaaniche, N., and Hammoudeh, M. (2019) Analysis of attribute-based cryptographic techniques and their application to protect cloud services. *Transactions on Emerging Telecommunications Technologies* 95: e3667.

4

eSIM-Based Authentication Protocol for UAV Remote Identification

Abdulhadi Shoufan[1], Chan Yeob Yeun[1], and Bilal Taha[2]

[1] *Center for Cyber-Physical Systems, Khalifa University, Abu Dhabi, United Arab Emirates*
[2] *Department of Electrical and Computer Engineering, University of Toronto, Toronto, ON, Canada*

4.1 Introduction

The market for unmanned aerial vehicles (UAV) or drones is growing rapidly. Diverse applications are profiting from this technology in construction, agriculture, insurance, the oil and gas industry, film making, sky photography, parcel delivery, journalism, security, law enforcement, and civil defense. The USA's Federal Aviation Administration (FAA), for example estimated that more than 450,000 nonmodel small drones were in operation in 2019. This sector is projected to have 835,000 aircraft in 2023 [1]. The global drone market is predicted to grow from $14 billion in 2018 to more than $43 billion in 2024 [2].

Many drones are equipped with various communication technologies (cellular systems [3], WiMAX [4], and satellite communication [5]) which enable them to an integral part of the Internet of Things (IoT) infrastructure. Drones can be equipped with visual sensors (cameras, visible light sensors, infrared sensors, etc.), mechanical sensors (e.g. pressure, movement, position, speed, acceleration, vibration, force, and momentum), weather sensors (temperature, humidity, and heat flow) as well as radiation, chemical, and acoustic sensors. These capabilities along with the ability to move freely in three dimensions have opened the door for several unique services [6].

Flying a drone, however, is knowingly associated with security, privacy, and safety risks which have shaped the progress and the penetration of this technology over the last years [7]. Safety is, without a doubt, the most urgent requirement when it comes to drone operation. This is confirmed by the frequent reports on drone incidents and intrusions worldwide. A chronological list of such incidents

Security and Privacy in the Internet of Things: Architectures, Techniques, and Applications,
First Edition. Edited by Ali Ismail Awad and Jemal Abawajy.
© 2022 The Institute of Electrical and Electronics Engineers, Inc. Published 2022 by John Wiley & Sons, Inc.

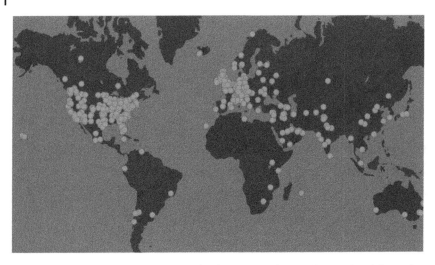

Figure 4.1 User interface to a web application that provides update-to-date information about drone incidents worldwide [8].

since September 2018 can be found online [8], see Figure 4.1. Security is another critical requirement for drone operations. This is because cyber and physical attacks on drones do not only threaten the security of information collected on the drone and exchanged with the ground station. Rather, a drone under security attacks such as control hijacking can be turned into a weapon against people, assets, and infrastructure [9]. The other way is also true: a drone, which violates safety regulations and is subject to increased risk of collision, crash, or interdiction, is exposed to a higher risk of losing secure information stored on the drone if malicious parties obtain physical access to the drone after crash or capture. Thus, safety and security are two complementary objectives for drone operations.

The link between drone security and safety can be highlighted in the context of remote identification. This technology describes the ability of a drone to provide its identity and other information in flight. The remote ID can be received by ground agents to verify several aspects such as the registration status of the drone. Also, law enforcement can use remote identification to find the controller when a drone appears in a no-fly zone or shows an unsafe flying manner. Furthermore, remote ID can help in situational awareness and air traffic management (ATM) contributing to the safety and security groundwork for sophisticated drone operations. The purpose of the remote ID can only be achieved if the identification message is shared in a secure way that prevents manipulation and falsification. So a ground agent should be able to verify that the received remote ID truly belongs to the drone transmitting it. Also, the remote ID message should be protected against manipulation attacks, e.g. to alter the actual location of the drone or its operator.

This chapter addresses the authentic communication of drones' remote identification and has the following contributions. First, the security of drones in the context of unmanned traffic management systems is investigated. This aspect is highly relevant for the controlled operation of commercial and civil drones today. Second, we highlight relevant security attacks on or from drones and show how a drone can be a target or a threat, respectively. The state-of-the-art in anti-drone technologies is then discussed including an overview of current detection, classification, and interdiction technologies. Toward the core contribution of the chapter, we then provide an in-depth review of the remote identification technology for drones. For this, we highlight relevant regulations and standards. The core contribution of the chapter is an authentication protocol for the secure communication of remote identification. This protocol relies on using an embedded Subscriber Identification Module (eSIM). A security analysis of the protocol is presented along with formal verification using ProVerif.

The remainder of the chapter is organized as follows. Existing methods on drone safe and secure operations are discussed in Sections 4.2 and 4.3. Then, Section 4.4 describes the UAV remote identification technology. In Section 4.5, the protocol design and its security analysis and formal verification are presented. The chapter is concluded in Section 4.6.

4.2 Drone Security

Related work has considered drones both as attack targets as well as security threats. A detailed review of related work on drone security can be found in [7]. An analysis of experts' assessment of drones' security objectives can be found in [10].

4.2.1 Drone Security in UTM

With the recent advances and large demands in several sectors such as e-commerce, surveillance, and agriculture, drones (also called small Unmanned Aircraft Systems [sUAS]) in different shapes and sizes are expected to fly around at low altitudes. These drones could be flying to monitor the traffic on streets, entertain families in their backyard, or deliver goods to urban areas. As a result, the FFA introduced the Unmanned Aircraft System Traffic Management (UTM) as a complementary traffic management system to the existing ATM system. The UTM system manages any uncontrolled drone operating under 400 feet in the airspace where the ATM is not supported [11]. Figure 4.2 shows a sample UTM structure in which small drones perform different tasks and operations while communicating with each other (U2U) or with the base infrastructure (U2I).

To ensure a safe as well as efficient UTM system, the security aspect should be investigated with more depth. Public security and safety are the top priority that

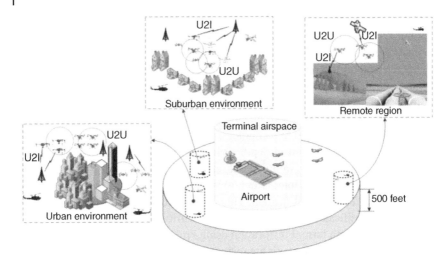

Figure 4.2 Illustration of some UTM systems and their operations as well as interactions in the airspace [11].

should be maintained at all times. Different types of security challenges can be found in UTM systems motivated by three categories of attack [11].

4.2.1.1 Physical Attacks

These attacks target the physical components of the UTM system or its security modules. The adversary uses tangible elements aiming at the UTM physical parts such as damaging the network foundation, stealing and sabotaging physical units of the UTM, forcefully overthrowing the small drones, and initiating a power outage to the UTM grid.

4.2.1.2 Cyber Attacks

These attacks aim at the cyber domain to damage or misuse the cyber components of the UTM system. The adversary in this case maliciously disables devices, steals or alters data, or starts a series of attacks by breaching the system computers. UTM-specific cyber threats include (i) The attacker can manipulate the perception of the time and space of the components. For example, the attacker can create the illusion of different physical locations of the drones or other objects in the system. (ii) The attacker can misuse the UTM system components for egotistical reasons such as manipulating a drone to attack a civilian or invade his/her privacy.

4.2.1.3 Cyber-Physical Attacks

These attacks breach the cyber domain of the system which results in a direct impact on the physical environment. The adversary aims to cause interference to the UTM system that would affect the physical asset's safety, performance as well as behavior. These attacks can target the network infrastructure, IoT controls, and aviation domain. Specific examples for the UTM system include attacks that modify the geofences or the ground data leading to higher flight risks for the small drones in the UTM system. Another example involves jamming the communication between the drones or between the drone and the station to block or corrupt the data exchange resulting in higher risks to the system. Figure 4.3 shows some common general cyber-attacks that can be performed on the UTM system. All the aforementioned attacks are specific to the UTM system, yet general security threats can also exist on drones. More details are exploited in Section 4.2.2.

4.2.2 Security Attacks on Drones

Several studies were presented which show successful attacks on drones either cyber or physical attacks. Figure 4.3 demonstrates common cyber and physical attacks on drones. The cyber-attacks can vary from simple man-in-the-loop threats targeting links in the system such as Command-and-Control (C2) as well as telemetry links to more sophisticated attacks such as spoofing a Global Navigation Satellite System (GNSS) receiver. These cyber-attacks can be seen as protocol-based attacks or sensor-based attacks.

Protocol-based attacks mainly focus on hijacking a drone by replaying messages sent from the ground station to the drone in the uplink. This was demonstrated in

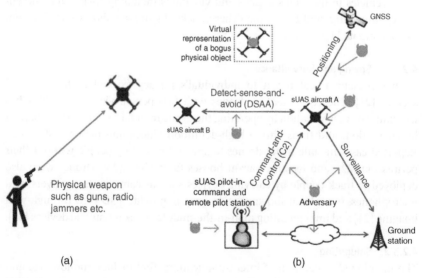

Figure 4.3 Illustration of possible attacks on the drones. (a) Physical attacks, (b) cyber-attacks [11].

[12] and [13] with attacks on the XBee 868LP protocol over 868 MHz frequency or the MAVLink protocol for amateur drones, respectively. Also, hijacking scenarios by stealing the control device were proposed and related solutions using behavioral biometrics were presented in [9, 14]. Sensor-based attacks, also known as spoofing attacks, target the drone's onboard devices such as the camera, the motion, and the pressure sensors. The authors in [15] demonstrated how to attack the gyroscope to force the drone to land. By spoofing the onboard camera using a laser and a projector, Davidson et al. demonstrated the possibility of causing stability issues in the drone [16]. The well-known GPS spoofing was also applied to hijack drones operating autonomous mode [17, 18].

Physical attacks comprise all attempts to bring down a drone by a malicious user [19]. Many interdiction technologies such as shooting using guns or capturing by other drones fall into this category. Furthermore, the usage of jammers to disrupt the GPS signal-causing drone malfunction [20] can be considered as a physical attack. Section 4.3.2 will describe such technologies in the context of anti-drone systems.

4.2.3 Security Attacks from Drones

The advancements in unmanned aircraft technologies made it possible to obtain low-price and low-size drones which can fly at high speeds up to 65 km/h and carry payloads of up to 6 kg. These features led to an increased deployment of these vehicles in malicious attacks and violations including spying and surveillance, smuggling, and launching cyber-attacks. Figure 4.4 shows samples from threats initiated by drones.

4.2.3.1 Spying and Surveillance
Drones present a real threat to individual's privacy as well as to industrial secrecy [21]. The equipment of drones with first-person view capabilities has facilitated privacy violation by operating the drone remotely, securing the channel by encryption, and transmitting high-definition videos and pictures of others. Reported cases include using drones to spy and recording people without their permission [22], and locating vacant houses for robbers [23]. Drones were also deployed to track people by carrying devices such as cameras, transceivers, and microphones. It has been shown that drones can provide a 3D view through-wall imaging [24] and implementing man-in-the-middle attack on mobile devices [25].

4.2.3.2 Smuggling
The mentioned integration of interesting features that include the low size and cost, the high speed as well as the ability to carry payload and to be controlled from a far distance has made drones an attractive tool for smugglers with two

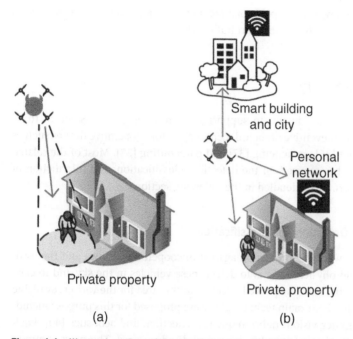

Private property

(a)

Private property

(b)

Figure 4.4 Illustration of possible attacks from drones. (a) Sample of privacy violation or physical attacks, (b) feasible cyber-attacks [11].

major advantages. First, the smuggling activities are getting less dependent on the capabilities of the individual smuggler. Second, the smuggler's identity can be kept unknown even if the drone is seized by law enforcement. Drones have been used to drop banned objects such as weapons into prison [26] and to smuggle drugs and goods across countries [27].

4.2.3.3 Physical Attacks
The term "terrorism by joystick" is referred to using drones by terrorists to carry out attacks. Examples of such attacks include the attempt to assassinate the Venezuelan president in 2018 [28] and the hostile attack at the USA's white house in 2015 [29]. In addition to that, drones are used on a larger scale, e.g. to explode in sensitive locations. Examples include threatening the Gatwick Airport [30].

4.2.3.4 Cyber Attacks
Several researchers demonstrated the ability of drones to execute cyber-attacks. Nassi et al. and Guri et al. showed how to penetrate and access communicated information by initiating a covert channel using drones with mounted transmitter and receiver, respectively [31, 32]. Other experiments illustrated that drones

with radio transceivers are capable of hijacking a Bluetooth mouse [33], a wireless printer [34], and a Philips Hue smart bulb [35].

4.3 Drone Safety

Different solutions are in use to support safe drone operation. Some of these solutions rely on preventive on-board technologies for cooperative drones such as Sense and Avoid [36], geofencing [37] and parachuting [38]. Most of the related work, however, has focused on the detection, classification, and interdiction of uncooperative drones as detailed in the following sections.

4.3.1 Drone Detection and Classification

The first step toward the protection against uncooperative drones and the associated cyber and physical risks is to detect these vehicles in the sky and to classify them. Machine learning proved effective in this area for the various available technologies [39]. Four main technologies were proposed for this purpose including radar, computer vision, radio-frequency detection, and acoustics [40]. Each of these technologies has its advantages and disadvantages. The most common system used for drone detection is the radar. Radars are resistant to noise and independent of line of sight (LOS). However, even with the variety of employed methodologies, it is not clear if the developed systems are capable of performing well with different drone types, wider ranges, and different radar sensors. The visual systems rely on cameras to detect the drones where LOS is needed. Most of the work utilizes deep learning to ensure high performance, yet, it is known that such models are data-driven which demands a lot of labeled data. On the other hand, acoustic systems do not require LOS, yet, it is very sensitive to ambient noise. Radio frequency detection systems are also independent of LOS, but they fail to detect drones flying autonomously. Table 4.1 provides a summary of the advantages and disadvantages of each technology. Research in this field should be taken to the next level by considering the limitations of each mentioned system and establishing a link with the industry. Besides, public datasets should be available with real-life consideration to develop high-performance and robust models.

4.3.2 Interdiction Technologies

A multitude of technologies has been developed to intercept, destroy, or capture small UAVs in urban airspace. Wyder et al. classified UAV interdiction technologies according to their impact on the target drone into three main categories,

Table 4.1 Comparison of advantages and disadvantages of different drone detection technologies.

Detection technology	Advantages	Disadvantages
Radar	Low-cost; less prone to noise compared to acoustic models; does not require a LOS compared to visual models; higher-frequency radars can capture the micro-Doppler signature.	Drones have small radar cross sections which make the detection more challenging; higher-frequency radars have higher path loss limiting their detection range.
Acoustic	LOS is not required; low-cost depending on the employed microphone arrays.	Sensitive to ambient noise especially in loud areas; needs a database of acoustic signatures for different drones for training and testing.
Visual	Low-cost depending on the utilized cameras and optical sensors; human assessment of detection results is easier using screens.	Visibility level affects the performance; expensive sensors such as thermal or laser-based may be needed; LOS is necessary.
Radio frequency	Low-cost RF-sensors; no LOS is required; long detection range.	Noneffective for autonomous drones without any communication channels; it requires training to learn RF signal signatures.

namely signal interception, propeller restriction, and aerial takedown [41]. Each of these technologies has key advantages and some drawbacks.

Due to its undisruptive nature, signal interception has received substantial attention in research and industry as a counter-drone solution in urban areas. According to [42], jamming-based interdiction can be divided depending on the mode of operation of the target drone into drone hacking and GPS spoofing. Fundamentally, the former is designed to target manually operated UAVs by breaking the communication link between the controlling pilot and the UAV via a deauthentication attack [43]. In contrast, the latter is targeted toward autonomously controlled UAVs where the flight maneuver is directed by a GPS signal. In this scenario, the GPS signal can be forged or jammed forcing the target UAV to change its flight path [44]. Despite being effective in eliminating potential attackers with minimum impact on the public, signal interception-based technologies are limited in certain aspects. Specifically, the introduction of GPS-denied localization and navigation modules has allowed drones to navigate independently of the GPS signal [41] rendering GPS spoofing ineffective. Moreover, new encryption methods of the UAV radio control signal limit the ability of hacking attacks on modern UAV platforms [44]. Most importantly, signal interception technologies are considered illegal as they interfere with cellular communication, radar, GPS, and other wireless networking services [45].

Alternatively, propeller restriction refers to capturing uncooperative drones through a capturing body which is usually a net. The net could be launched either manually by a skilled operator on the ground [46], or autonomously by another flying drone [47]. Although this approach is the most popular, it is limited by two factors: (i) the range at which the hostile drone can be captured, and (ii) the skills of the ground gun operator in case it is manually operated.

A variety of aerial takedown technologies are available in the military and civilian counter-drone literature. These include (i) hunting by eagles [48], (ii) shooting by machine guns or laser [42], (iii) control signal jamming [49]. All these counter-drone techniques result in forcing the UAV to land or to be destroyed. Although proven most efficient, takedown methods are the most disruptive and unsafe when it comes to public safety as they usually cause the target drone to descend uncontrollably.

4.4 UAV Remote Identification

Aviation authorities worldwide are mandating the deployment of remote identification to promote safe and accountable drone operations. The European Union Aviation Safety Agency (EASA) has published related regulations in March 2019 (Commission delegated regulation (EU) 2019/945) that were amended in April 2020 ((EU) 2020/1058). These regulations mandate that all unmanned aircraft should be equipped with a remote identification system [50]. The FAA in the USA has published a final rule for remote identification in January 2021 [51]. The document provides regulations for three types of remote identification:

1. *Standard Remote ID*: The drone uses a built-in capability to broadcast identification and location information about the drone and its control station. Drones with standard remote ID can operate beyond visual line of sight (BVLOS).
2. *Remote ID broadcast module*: The drone uses an add-on device to broadcast identification and location information. This option is provided to retrofit currently available drones with remote ID capability. However, unmanned aircraft using this type of remote ID must operate in the pilot's LOS.
3. *FAA-recognized identification areas*: Hobbyists, as well as educational or research institutions, can fly drones in specific approved areas with the need to broadcast remote identification (Figure 4.5).

Remote ID regulations defined by aviation authorities are typically performance-based without exact specification of supporting technologies. Instead, these regulations frequently refer to technical standards as possible ways of compliance.

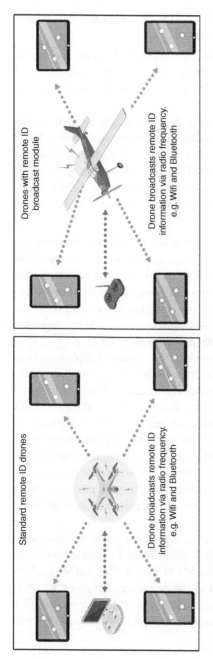

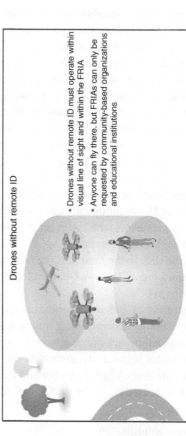

Standard remote ID drones

Drone broadcasts remote ID information via radio frequency. e.g. Wifi and Bluetooth

Drones without remote ID

- Drones without remote ID must operate within visual line of sight and within the FRIA
- Anyone can fly there, but FRIAs can only be requested by community-based organizations and educational institutions

Drones with remote ID broadcast module

Drone broadcasts remote ID information via radio frequency. e.g. Wifi and Bluetooth

Figure 4.5 Drone pilots have three ways to meet the FAA's Remote ID rule Source: Federal Aviation Administration [51].

One of such standards is the American Society for Testing and Materials (ASTM) Standard Specification for Remote ID and Tracking [52]. This standard defines two technologies for drones' remote identification:

1. *Network-based remote identification*: This method helps to make the remote identification information available globally via the Internet on dedicated servers. The servers can be queried by observers for different purposes including situational awareness.
2. *Broadcast-based remote identification*: Using this method, the drone transmits its identification locally using one-way communication over Bluetooth or Wi-Fi without using the Internet protocol between the data link and application layers. A local observer can receive the remote ID in real-time using any handheld device that supports the proposed communication links. The Wi-Fi Neighbor Awareness Networking is proposed to support broadcast-based remote identification [53].

Network-based identification depends on Internet connectivity between the drone and the observer through Unmanned Service Suppliers (USS) servers, from the UAS to each Observer. Broadcast remote ID (RID) should need Internet (or other Wide Area Network) connectivity only for UAS registry information lookup using the directly locally received Unmanned Aircraft System Identifier (UAS ID) as a key. Broadcast RID does not assume IP connectivity of UAS; messages are encapsulated by the UAS without IP, directly in Bluetooth or Wi-Fi.

The main motivation behind using network-based remote identification is to integrate drones into UTM systems and to facilitate the introduction of global identification for drones. In the notice of proposed rulemaking (NPRM) announced in December 2019, the FAA originally required that drones should support both network-based as well as broadcast-based identification [54]. The FAA received more than 53,000 comments from the public and many of these comments referred to privacy concerns with the network-based solution. This is because this identification method requires the storage of relevant data on intermediate servers belonging to the USS. In the final rule, the FAA responded to these concerns by removing the requirement of network-based identification and keeping the broadcast-based solution [51].

Regardless of the used communication technique, the identification information should be embedded into a RIDM that should be sent periodically. Most regulations specify one second for the RIDM period. The ASTM standard differentiates between dynamic data such as the drone location and static data such as the drone's serial number and assigns a longer period (three seconds) for the latter. According to the FAA's final remote identification rule, the RIDM has the elements:

1. Drone identification
2. Drone location and altitude

3. Drone velocity
4. Control station location and elevation
5. Time mark
6. Emergency status

According to the ASTM standard, drone identification can be one of three types [52]:

1. *Type-1*: This is a static hardware serial number assigned by the manufacturer according to the ANSI/CTA-2063-A standard for Small Unmanned Aerial System Serial Numbers [55].
2. *Type-2*: This is also a static ID assigned by an aviation authority similar to the registration number of a manned aircraft.
3. *Type-3*: This ID is assigned by a UTM system and can be static or dynamic, whereas the latter is sometimes referred to as session ID. The session ID is an alphanumeric code that is generated randomly by a Remote ID USS and assigned to the drone on a per-flight basis. The goal of the session ID is to enhance the privacy of the operator and the secrecy of the mission.

The EASA regulations in Europe ((EU) 2020/1058) allow Type-1 only [50]. In the USA, the proposed rulemaking supported Type-1 and Type-3 IDs [54]. However, with the removal of network-based identification in the final rule, the application scope of Type-3 IDs is still to be determined [51]. Type-2 ID does not seem to have been adopted in any regulations, so far. The FAA explicitly requires that the remote ID does not include the drone registration number and that only the FAA and selected agencies such as law enforcement are authorized to use information in the remote ID to look up the registration number. This indicates that session IDs would be linked to the registration numbers indirectly through the serial number in the registration database.

Regulation authorities, standardization institutions, and industry all highlight the importance of the cyber-physical security of remote identification. For example, tamper-resistant hardware is proposed to hinder unauthorized change of the remote identification equipment. Furthermore, cybersecurity protections for the transmission and broadcast of the message elements should be incorporated to defend against cyber threats that could adversely affect the authenticity or integrity of the remote identification information. In the next section, we present a solution for remote ID authentication.

4.5 Authentication Protocol for Remote Identification

Researchers have proposed various protocols for establishing secure communication with a focus on individual security services such as confidentiality, integrity, availability, and authentication. In this section, we introduce a security protocol for remote identification of drones following two objectives: (i) The protocol should cover multiple security services including confidentiality, authentication, availability, and integrity. (ii) Communication and information sharing between the smartphone and the UAV are limited to authorized users. We first describe the preliminaries and assumptions for the protocol design. Then, we define the protocol components and their roles. After that, we explain the proposed protocol and implementation scenarios. Finally, we provide a security analysis of this protocol.

4.5.1 Preliminaries

The proposed protocol design is based on the assumption that the drone is equipped with an *eSIM* and that the user's smartphone communicates with the drone using this eSIM as illustrated in Figure 4.6. eSIM is an unremovable hardware module, which supports multiple mobile operator profiles. The switch from one profile to other works without replacing the card as known in the traditional SIM [56]. Figure 4.7 demonstrates the operation of the eSIM service.

4.5.1.1 Assumptions and Notations

The system has the following agents:

1. Operator smartphone including a secure application to communicate with the UAV as a remote controller.
2. UAV incorporating an eSIM service.
3. Mobile Network Operator (MNO).
4. Government Telecommunication Authority (GTA).
5. Verifier smartphone. A verifier is an agent who is interested in receiving a drone's remote ID and verifying the authenticity of this ID.

The proposed security protocol relies on the following assumptions:

1. The eSIM embedded into the vehicle is developed by the GTA.
2. The secure application is developed by the GTA.
3. The MNO communicates with the GTA in the background.
4. A port-hopping technique is available for secure communication between the operator smartphone and the UAV.

Figure 4.6 A UAV with embedded SIM (eSIM) and operator smartphone.

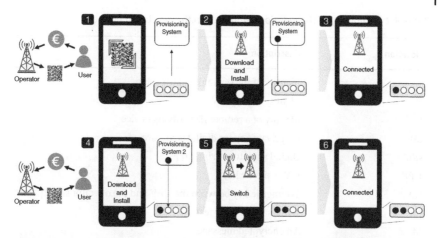

Figure 4.7 In Step (1), the user must scan the Quick Response code which allows the UAV to connect to the system and securely install the eSIM profile as it shows in Step (2). Then in Step (3), the user will be able to connect to the provided network operator. If the user wishes to use a different operator, he or she sets up a contract with the second operator as it shows in Step (4) and scans the Quick Response code to download the new profile. In Step (5) and (6) the user will be able to switch between the profiles and change the operator's network Source: Pannell and eSIM Whitepaper [57].

5. The system supports the public key infrastructure and X.509 certificates.
6. Advanced Encryption Standard is used for symmetric-key encryption.
7. Elliptic Curve Cryptography (ECC) is used for asymmetric-key encryption.
8. The eSIM subscription has a limited number of users.
9. The application on the operator smartphone and the vehicle share the secret PIN at the registration phase.
10. The GTA stores information about each eSIM subscriber in a database.

Port hopping "is a typical proactive cyber defense technology, which hides the service identity and confuses attackers during reconnaissance by constantly altering service ports" [58]. Using this technique helps to prevent Denial-of-Service (DOS) and Distributed Denial-of-Service (DDOS) attacks. X.509 certificates are generated by the GTA and used to verify the authenticity of a public key. A X.509 certificate contains (i) Information about the owner, issuer, and the certificate validity date. (ii) Public key of the owner and the cryptographic algorithm. (iii) Signatures from the issuing Certificate Authority [59]. We use ECC instead of Rivest–Shamir–Adleman for asymmetric encryption because the former can support the same security level with a smaller key. Table 4.2 summarizes the notations used in the protocol.

Table 4.2 Summary of the notations employed in the developed protocol.

Notation	Description
A	Identity of the operator smartphone
B	Identity of the UAV
C	Identity of a remote ID verifying device
SB	B's private key for digital signature
sSB	Digital signature using SB
KPB	UAV's public key
KAB	Symmetric key between the UAV and the operator smartphone
m	A message in clear text
M	An encrypted message
CC	Control command from operator to UAV
CC'	UAV's response to a control command
e	Encryption
RIDM	Remote Identification Message
Port #	Port hoping number between smartphone and UAV
RA1, RA2	NONCEs generated by operator's smartphone
RB1, RB2	NONCEs generated by the UAV
TB	Time stamp
Hash1	Hash of the message 1 generated by smartphone
x||y	Concatenation of data items x and y

4.5.2 Registration

Initially, the user must register to the eSIM service by sending a request to the MNO. As mentioned earlier, we assume that this technology is already embedded into the UAV by the GTA [60]. Then, the user visits the MNO office physically to provide information about the UAV. After verifying the user information, the MNO communicates with the GTA requesting the PIN and the activation code that will be shared with the user to complete the registration. The MNO requests the user to install and use the secure application that will manage the UAV connection. Recall that this application is developed by GTA to allow secure communication between the operator smartphone and the eSIM-equipped drone, see Figure 4.8.

In the second phase illustrated in Figure 4.9, the user runs the application to activate the eSIM and install the profile in the vehicle using the activation code

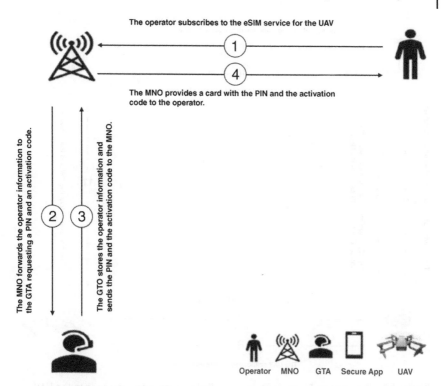

Figure 4.8 Registration Phase 1: obtaining the PIN and the activation code for the eSIM service.

and the PIN as well as the last seven characters of the vehicle's serial number. The application sends the PIN and the activation code secretly to the GTA, which stores the user and the vehicle information in the database. After that, the GTA pushes the PIN code in the vehicle which has the same last seven characters of the UAV serial number. The UAV confirms its readiness to receive other information from the user itself. The GTA sends a notification to the user through the application. The user locally enters the activation code and the PIN code in the vehicle. At this stage, both the drone and the operator are sharing the secret PIN and the activation code which allows them to identify each other.

4.5.3 Secure Communication Protocol

After completing the registration process, the operator can communicate securely with the UAV using the proposed protocol shown in Figure 4.10. The exchanged messages M1, M2, M3, and M4 will be explained in the following.

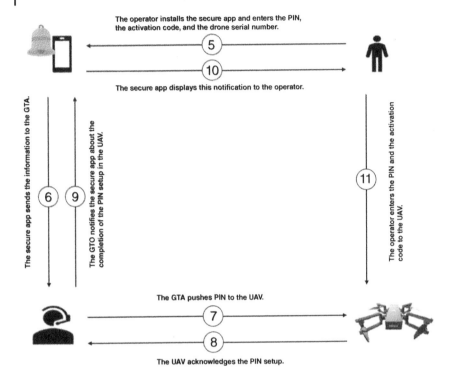

Figure 4.9 Registration Phase 2: flashing the PIN and the activation code to the UAV.

Figure 4.10 The proposed protocol for secure communication between the operator's control device and the UAV.

4.5.3.1 M1: A Challenge from the Operator (A) to the Drone (B)

Initially, we assume that port 444 is enabled for communication between the operator and the drone. M1 is generated by the operator and consists of two concatenated parts:

1. A message $m1 = (RA1 \| B \| KAB \| Port\ 110 \| PIN\ code \| RIDM)$ encrypted with the drone's public key (KPB).
2. The hash value of m1 which is *Hash1*.

$m1$ includes the nonce of the operator smartphone (RA1), the identity of the drone (B), the session Key (KAB), the next port to be used (Port 110), the PIN code obtained in the registration, and those segments of the remote identification message related to the operator, namely his location and latitude (remote identification message (RMID) [RMID]).

Upon receiving $M1$, the drone decrypts the first part using its drone private key (KB) to obtain $m1$. The drone then hashes $m1$ and compares the calculated hash value with the received one *Hash1*. If both values are identical, then the drone assumes that the message was not modified en route. Then the drone checks if the nonce RA1 was used before. If this is not the case, the drone assumes that the message is fresh, i.e. there is no replay attack. Finally, the drone compares the received PIN code with the one the drone has obtained in the registration. If both are identical, then the message is assumed to stem from the authentic operator.

4.5.3.2 M2: A Response from the Drone (B) to the Operator (A)

The drone constructs a message $m2 = (RB1 \ || \ RA1 \ || \ A \ || \ Port \ 112 \ || \ RIDM)$. It consists of a nonce generated on the drone (RB1), the nonce of the operator smartphone (RA1), the identity of the operator (A), the next port to be used (Port 112), and those segments of the remote identification message related to the drone, namely its serial number, location, latitude, velocity, and emergency status (RIDM). $m2$ is then encrypted with the shared session key (KAB) which is retrieved from M1.

Upon receiving M2, the operator decrypts it using the shared session key (KAB). The operator retrieves RA1 and compares it with the sent one. If both are identical, then the operator is assured that M2 is coming from his drone since no one else knows this nonce.

4.5.3.3 M3: Control Message from the Operator to the Drone

After establishing a secure connection between the operator and the drone through the challenge (M1) and the response (M2), the operator can now control the drone using this secure connection. M3 is a typical control message which is encrypted with the shared session key (KAB). The plain text message $m3 = (RA2 \ || \ RB1 \ || \ B \ || \ Port \ 114 \ || \ CC \ || \ RIDM)$ contains a fresh nonce from the operator (RA2), the previous drone's nonce (RB1), the drone serial number (B), the port for the next message (Port 114), the control command (CC), and an updated remote identification message RIDM.

Upon receiving M3, the drone decrypts it using the shared session key (KAB), retrieves RB1, and verifies its correctness to conclude about the authenticity of the control message. The CC is decoded and executed. Depending on the operation mode of the drone and its service, this command – for example – can include data for a new position the drone should fly to or a command to capture an image, etc. Also, the drone updates the remote identification message according to the RIDM segment in $m3$.

4.5.3.4 M4: Drone's Response to the Control Message

The drone responds to the control message by encrypting $m4 = (CC' \parallel RIDM \parallel RB2 \parallel RA2 \parallel A)$ with KAB. This includes a fresh drone's nonce RA2 and an updated version of the remote identification data. Also, a response to the CC can be sent (CC'). For example, this can be an image that the operators asked for. The RIDM in $m4$ includes updated information about the drone location, latitude, velocity, and emergency status.

4.5.3.5 M5: Secure Broadcast of Remote Identification Message

Periodically, the drone sets up a remote identification message containing information about the drone itself as well as its operator according to a specified message format. Regulations require the inclusion of a time mark, which essentially indicates the time point at which sensor readings occurred. Since the RMID has operator-related information, we propose to include another timestamp (TB) to the remote identification message for security purposes.

Figure 4.11 demonstrates the secure broadcasting message from the drone to any potential verifier device. The message contains RIDM concatenated with the timestamp TB as well as the digital signal of the hash value of the segment TB $\parallel$ RMID. The digital signature is determined using the drone's private key. Upon receiving this message, the verifier first calculates the hash value of the received segment TB $\parallel$ RMID. Then it uses the drone's public key (KPB) to verify the signature. The outcome of this verification is compared with the calculated hash value to verify the authenticity of the message.

4.5.4 Security Analysis

We analyze our protocol based on the four requirements discussed in [61]:

1. Are the messages sent by the required sender confidentially?
2. Are the messages fresh to prevent replay attacks?
3. Are the messages intended for the right receiver?
4. Are the messages in sequence?

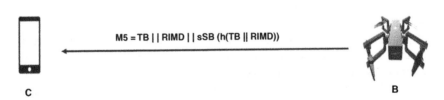

$$M5 = TB \parallel RIMD \parallel sSB\ (h(TB \parallel RIMD))$$

C

B

Figure 4.11 Secure broadcasting of the remote identification message.

In the following, we will verify these requirements for the exchanged message starting with M1 and M3. These two messages are repeated here:

M1 = eKPB ((RA1 || B || KAB || Port 110 || PIN code) || Hash1)
M3 = ekAB (R′A2 || RB1 ||B || Port 114 || CC || RIDM)

1. *Can the drone make sure that M1 and M3 were generated by their legitimate operator?* As M1 contains the secret "PIN code" the drone can verify that this message stems from its operator because this code is only known to these two parties through the registration procedure. Also, the drone can verify that M3 comes from its operator because of the encrypted RB1 which is only known to these two parties with a key (KAB) known to them only, too.
2. *Can the drone make sure that M1 and M3 are fresh and not replays of old messages?* Yes, this is guaranteed by the numbers used once RA1 in M1 and R′A2 in M3.
3. *Can the drone make sure that M1 and M3 were intended for the drone itself?* Yes, because of the usage of the drone identity (B) in both messages.
4. *Can the drone make sure that M1 and M3 were generated in the right sequence?* Yes, because of the inclusion of the nonces RA1, RB1, and R′A2.

Now, we verify how the four requirements apply for messages M2 and M4:

M2 = eKAB (RB1 || RA1 || A || Port 112 || RIDM)
M4 = eKAB (CC′ || RIDM || R′ B2 || R′ A2 || A)

1. *Can the operator make sure that M2 and M4 were generated by his drone?* Yes, because these messages are encrypted with KAB.
2. *Can the operators make sure that M2 and M4 are fresh and not replays of old messages?* Yes, because of using the nonce RB1 in M2 and R'B2 in M4.
3. *Can the operator make sure that M2 and M4 were intended for himself?* Yes, because of using the operator identity (A) in both messages.
4. *Can the operator make sure that M2 and M4 were generated in the right sequence?* Yes, because of the inclusion of RA1, RB1, R'A2, and R'B2.

Finally, we verify these requirements for M5:

1. *Can the verifier be sure that M5 was generated by the drone?* Yes, because of using the signature function which provides evidence for the origin of the message.
2. *Can the verifier be sure that M5 is fresh?* Yes, given that TB is sufficiently recent and no similar message has been received in the current time window.
3. *Can the verifier be sure that M5 was intended for himself?* Not applicable because M5 is a broadcasting message.
4. *Can the verifier make sure that M5 was generated in the right sequence?* This requirement is not applicable here due to the broadcasting nature of this message.

4.5.5 Formal Verification

In this section, we present the formal verification of the proposed protocol to prove its secrecy and mutual authentication. Analyzing security protocols by a human is knowingly difficult. Many tools such as ProVerif, Scyther, Athena, cryptographic protocol shapes analyzer (CPSA), and automated validation of Internet security-sensitive protocols and applications (AVISPA) are available to support this task. We decided to use ProVerif (version 2.00) because it does not require the definition of intruders and attackers, the specification of communication channels, or any special code for freshness attacks [62]. Furthermore, ProVerif supports several cryptographic primitives beyond digital signatures, symmetric and asymmetric encryption, and hash functions. ProVerif can handle input files written in the pi-calculus language and can prove security properties such as secrecy and authentication of security protocols. This tool was used successfully to test several known protocols such as the "Needham-Schroeder" [63,64]. ProVerif shows several useful properties including reachability, correspondence assertions, and observational equivalence. Reachability refers to the investigator's ability to see how the protocol terms are available to the attacker. Correspondence assertions allow studying the relationships between the events. Observational equivalence means that the adversary cannot distinguish between two processes. Table 4.3 provides a summary of the threats/attacks that should be prevented by our protocol and mechanisms of this protection.

4.5.5.1 Declaration of User-Defined Types and Terms

Now we describe the implementation of the protocol step by step using ProVerif. Table 4.4 summarizes some custom types we defined and related names that are used in the cryptographic primitives in our protocol. The syntax for declaring a user-defined type is

type s.
s is the type name.

Table 4.3 Threats and countermeasures.

Threats	Countermeasures
Impersonation	Identity-based mutual authentication and using a secret PIN
Replay attack	Freshness checking using NONCE
Denial of service	Port hoping
Man-in-the-middle attack	Mutual authentication
Sharing session key	Use session keys to encrypt the rest of the message and including the PIN code
Nonrepudiation	Including the PIN code

Table 4.4 User-defined types and declared terms.

Type	Description	Terms
skey	UAV private key	Kb
pkey	UAV public key	pkeyB
keys	Session symmetric key	Kab
smartdevice	An entity	A
vehicle	Another entity	B
nonce	Number used once	n1, n2, n3, n4
portN	Hoping port	p1, p2, p3
m	Request or response message	ReqA, ResB
pincode	PIN between UAV and smartdevice	Pin

Free names are declared using the syntax:

free n: t.

n is the name, **t** is the type, and **free** indicates that the name is available to all processes [63].

Free names that are private to the attacker are declared using the syntax:

free n:s [private].

We define a channel that is used for public communication. An adversary or attacker has complete control of the channel and can read, delete, modify, or even inject messages into this channel.

free c: channel.

4.5.5.2 Declaration of Cryptographic Primitives

Cryptographic primitives are modeled using constructors in ProVerif. A constructor is declared using the syntax:

fun f(t1,…, tn): t. f is a constructor of arity **n**. **t1** to **tn** are the types of the arguments and **t** is the return type. Constructors are available to the attacker unless they are declared private:

fun f(t1,…, tn): t [private].

Private constructors can be used to model tables of keys stored by the server, for example. The relationships between cryptographic primitives are captured by destructors that are used to manipulate terms formed by constructors. Destructors are modeled using rewrite rules given below form, where g is a destructor of arity k. The terms $M_{1,1}, …, M_{1,k}, M_{1,0}$ result from applying the constructors to the terms $x1,1, …, x_{1,n_1}$ of types $t_{1,1}, …, t_{1,n_1}$, respectively.

reduc for all $x_{1,1}: t_{1,1}, ..., x_{1,n_1}: t_{1,n_1}; g(M_{1,1}, ..., M_{1,k}) = M_{1,0};$

...

for all $x_{m,1}: t_{m,1}, ..., x_{m,n_m}: t_{m,n_m}; g(M_{m,1}, ..., M_{m,k}) = M_{m,0};$

4.5.5.3 Examples

The following code shows the declaration of a primitive for symmetric-key encryption. In the following code, the first line declares a symmetric-key encryption function. The second line declares its destructor which enables the decryption. m is the message which contains the message sent by the smartdevice, and k_2 is the symmetric key.

fun senc(bitstring, keys): bitstring.
reduc forall m: **bitstring,** k_2: **keys; sdec(senc(**m, k_2)**,**k_2)=m.

The following code shows the declaration of a primitive for asymmetric-key encryption. In the first line, the constructor **pk** takes an argument of type **skey** (private key) and returns a **pkey** (public key). The second line declares the asymmetric encryption function using the public key. The third line declares its destructor which enables the decryption. m is the message which contains the message sent by the smartdevice, and the k_2 is the symmetric key.

fun pk(skey): pkey.
fun aenc(bitstring, pkey): bitstring.
reduc forall m: **bitstring,** k_1: **skey; adec(aenc(**m, **pk(**k_1)**),**k_1)=m.

The code below shows the declaration of a hashing primitive. Here, we define the message which contains the private and public values (nonce, smartdevice, vehicle, keys, portN, pincode) to be hashed.

fun h(nonce, smartdevice, vehicle, keys, portN, pincode):bitstring.

4.5.5.4 Reachability and Secrecy Checking

To prove the reachability and secrecy in ProVerif, we investigate which terms are available to an attacker using the syntax:

query attacker (M).

We verify the secrecy of the session keys, nonce, port numbers, the UAV private key, PIN code, the request message, and the response message. Recall that all these terms were declared at the beginning of the protocol as seen in Table 4.4:

query attacker (Kab).
query attacker (n1).
query attacker (n2).
query attacker (n3).
query attacker (n4).
query attacker (Kb).
query attacker (p1).
query attacker (p2).

query attacker (p3).
query attacker (ReqA).
query attacker (ResB).
query attacker (Pin).

ProVerif uses the Horn theory to search for a security gaps in the protocol. Executing ProVerif produced the following output with respect to these queries which prove the secrecy of the respective information:

Query not attacker (kab) true
Query not attacker (n1) true
Query not attacker (n2) true
Query not attacker (n3) true
Query not attacker (n4) true
Query not attacker (kb) true
Query not attacker (p1) true
Query not attacker (p2) true
Query not attacker (p3) true
Query not attacker (ReqA) true
Query not attacker (ResB) true
Query not attacker (Pin) true

4.5.5.5 Verifying Mutual Authentication

ProVerif can be used to verify the mutual authentication between the smart device (A) and UAV (B) through using the event technique. Figure 4.12 illustrates eight events that are needed to achieve the mutual authentication between A and B.

- Event 1: A request is sent from A to create a trusted session with B.
- Event 2: B accepts the trusted session process.
- Event 3: Response from B.
- Event 4: A starts a trusted session.

After sharing the secret key between A and B, the next events are used for application purposes:

- Event 5: A sends a request to B asking for a pickup at a specific time and location.
- Event 6: B receives and accepts the request.
- Event 7: B sends the response, which clarifies the estimated time to arrive and the traffic state.
- Event 8: A receives the response and accepts it.

The below events and queries are defining the mutual authentication of our proposed protocol using the ProVerif tool, whereas

- Event1 (BeginAB()) and Event2(BeginBA()) are defined as event E1.
- Event3 (EndBA()) and Event4 (EndAB()) are defined as event E2

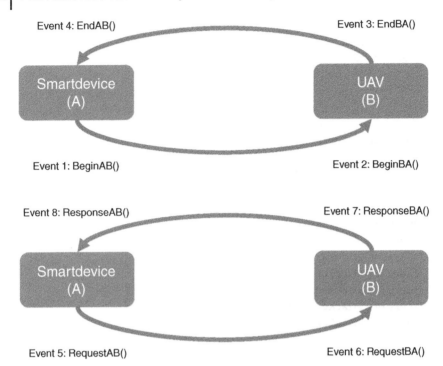

Figure 4.12 Eight events for testing mutual authentication in the proposed protocol.

- Event5 (RequestAB()) and Event6 (RequestBA()) are defined as event E3.
- Event7 (ResponseBA()) and Event8 (ResponseAB()) are defined as event E4.

  event E1 (nonce, portN, keys, pincode).
  event E2 (nonce, nonce, portN).
  event E3 (nonce, nonce, portN, M).
  event E4 (nonce, nonce, M).
  query n1:nonce, n2:nonce, p1:portN, kab1:keys, pin:pincode; event (E2(n1,n2,p1)) ==> event (E1(n1, p1, kab1, pin)).

  query n2:nonce, n3:nonce, n4:nonce, p3:portN, ResB1:M, ReqA1:M; event (E4(n4,n3, ResB1)) ==> event (E3(n3,n2,p3,ReqA1)).

  query n4:nonce, n1:nonce, n3:nonce, ResB2:M, p1:portN, pin:pincode, kab1:keys; event (E4(n4,n3,ResB2)) ==> event (E1(n1,p1,kab1,pin)).

  query n4:nonce, n1:nonce, n3:nonce, ResB2:M, p1:portN, n2:nonce; event (E4(n4,n3,ResB2)) ==> event (E2(n1,n2,p1)).

  The outcome of this testing of mutual authentication is given below which shows

  Query event (E2()) ==> event(E1()) is true.
  Query event (E4()) ==> event(E3()) is true.

Query event (E4()) ==> event(E1()) is true.
Query event (E4()) ==> event(E2()) is true.

The ProVerif tool is not able to prevent DOS attacks. However, by implementing the port hoping technique, the protocol can prevent or mitigate the risk of DOS/DDOS attacks. Finally, the encryption and decryption in the proposed protocol need two processes A and B. These two processes execute multiple times in parallel. The syntax below represents the process:

process
(
(!ProcessA) | (!ProcessB)|
)

4.6 Conclusion

The drone technology has shown success in diverse applications and it promises an expanded range of useful, effective, and efficient services. Opening the airspace to drones, however, will increasingly invite malicious users who will try to violate the rules by operating their drones without authorization. Remote identification is a promising technology that will allow the identification of cooperative drones. However, this technology is inherently not secure and we need to make sure that a received remote ID is an original one. This chapter presented a security protocol that allows any user to verify the authenticity of a remote ID by using digital signatures. This solution assumes that every drone has a public key that can be accessed by any interested verifier. The infrastructure required for this purpose was not discussed in this chapter and will be addressed in future work.

References

1 FAA (2019). Unmanned Aircraft Systems. https://www.faa.gov/data_research/ aviation/aerospace:forecasts/media/\LY1\textbackslashunmanned_aircraft_ systems.pdf (accessed 20 September 2021).

2 DroneII (2019). The Drone Market 2019-2024: 5 Things You Need to Know. https://www.droneii.com/the-drone-market-2019-2024-5-things-you-need-to-know (accessed 20 September 2021).

3 Qazi, S., Siddiqui, A.S., and Wagan, A.I. (2015). UAV Based real time video surveillance over 4G LTE. *2015 International Conference on Open Source Systems & Technologies (ICOSST)*, pp. 141–145. IEEE.

4 Rahman, Md.A. (2014). Enabling drone communications with WiMAX technology. *IISA 2014, The 5th International Conference on Information, Intelligence, Systems and Applications*, pp. 323–328. IEEE.

5 Skinnemoen, H. (2014). UAV & Satellite communications live mission-critical visual data. *2014 IEEE International Conference on Aerospace Electronics and Remote Sensing Technology*, pp. 12–19. IEEE.

6 Motlagh, N.H., Taleb, T., and Arouk, O. (2016). Low-altitude unmanned aerial vehicles-based internet of things services: comprehensive survey and future perspectives. *IEEE Internet of Things Journal* 3 (6): 899–922.

7 Altawy, R. and Youssef, A.M. (2016). Security, privacy, and safety aspects of civilian drones: a survey. *ACM Transactions on Cyber-Physical Systems* 1 (2): 1–25.

8 DeDrone (2020). Worldwide Drone Incidents. https://www.dedrone.com/resources/incidents/all (accessed 20 September 2021).

9 Shoufan, A., Al-Angari, H.M., Sheikh, M.F.A., and Damiani, E. (2018). Drone pilot identification by classifying radio-control signals. *IEEE Transactions on Information Forensics and Security* 13 (10): 2439–2447.

10 Shoufan, A. and Damiani, E. (2017). On inter-rater reliability of information security experts. *Journal of Information Security and Applications* 37: 101–111.

11 Sampigethaya, K., Kopardekar, P., and Davis, J. (2018). Cyber security of unmanned aircraft system traffic management (UTM). *2018 Integrated Communications, Navigation, Surveillance Conference (ICNS)*, pp. 1C1–1. IEEE.

12 Rodday, N. (2016). Hacking a Professional Drone. Slides at www.blackhat.com/docs/asia-16/materials/asia-16-Rodday-Hacking-A-Professional-Drone.pdf (accessed 20 September 2021).

13 Highnam, K., Angstadt, K., Leach, K. et al. (2016). An uncrewed aerial vehicle attack scenario and trustworthy repair architecture. *2016 46th Annual IEEE/IFIP International Conference on Dependable Systems and Networks Workshop (DSN-W)*, pp. 222–225. IEEE.

14 Shoufan, A. (2017). Continuous authentication of UAV flight command data using behaviometrics. *2017 IFIP/IEEE International Conference on Very Large Scale Integration (VLSI-SoC)*, pp. 1–6. IEEE.

15 Son, Y., Shin, H., Kim, D. et al. (2015). Rocking drones with intentional sound noise on gyroscopic sensors. *24th {USENIX} Security Symposium ({USENIX} Security 15)*, pp. 881–896.

16 Davidson, D., Wu, H., Jellinek, R. et al. (2016). Controlling UAVs with sensor input spoofing attacks. *10th USENIX Workshop on Offensive Technologies (WOOT) 16)*.

17 Noh, J., Kwon, Y., Son, Y. et al. (2019). Tractor beam: safe-hijacking of consumer drones with adaptive GPS spoofing. *ACM Transactions on Privacy and Security (TOPS)* 22 (2): 1–26.

18 He, D., Qiao, Y., Chen, S. et al. (2018). A friendly and low-cost technique for capturing non-cooperative civilian unmanned aerial vehicles. *IEEE Network* 33 (2): 146–151.

19 FORTEM Technologies. DroneHunter: Net Gun Drone Capture: Products. https://fortemtech.com/products/dronehunter/ (accessed 20 September 2021).

20 Robinson, M. (2015). Knocking my Neighbor's Kid's Cruddy drone offline. *DEF CON* 23.

21 Grubesic, T.H. and Nelson, J.R. (2020). Small unmanned aerial systems and privacy. In: *UAVs and Urban Spatial Analysis*, 71–87. Springer. https://doi.org/10.1007/978-3-030-35865-5.

22 Evans, A. For Mailonline (2017). Woman Grabs Gun and Shoots at Nosy Neighbour's Drone. https://www.dailymail.co.uk/news/article-4283486/ Woman-grabs-gun-shoots-nosy-neighbour-s-drone.html (accessed 20 September 2021).

23 Barrett, D. (2015). Burglars Use Drone Helicopters to Target Homes. https:// www.telegraph.co.uk/news/uknews/crime/11613568/Burglars-use-drone-helicopters-to-identify-targe-homes.html (accessed 20 September 2021).

24 Karanam, C.R. and Mostofi, Y. (2017). 3D Through-wall imaging with unmanned aerial vehicles using WiFi. *2017 16th ACM/IEEE International Conference on Information Processing in Sensor Networks (IPSN)*, pp. 131–142, April 2017.

25 Vemi, S.G. and Panchev, C. (2015). Vulnerability testing of wireless access points using unmanned aerial vehicles (UAV). *Proceedings of the European Conference on e-Learning*, Volume 245. Academic Conferences and Publishing International.

26 Davis, D. (2019). Correctional Officials Raise Concern over Drones Smuggling Contraband into Kingston-Area Prisons, Mar 2019. https://globalnews.ca/news/ 5074018/drones-smuggling-contraband-into-prisons-kingston/ (accessed 20 September 2021).

27 South China Morning Post (2019). Police Smash Smuggling Operation that Used Drone to Send Goods to Mainland, Apr 2019. https://www.scmp.com/ news/china/society/article/3006919/police-break-smuggling-operation-used-drone-send-us74566-goods (accessed 20 September 2021).

28 Herrero, A.V. and Casey, N. (2018). Venezuelan President Targeted by Drone Attack, Officials Say, Aug 2018. https://www.nytimes.com/2018/08/04/world/ americas/venezuelan-president-targeted-in-attack-attempt-minister-says.html (accessed 20 September 2021).

29 Schmidt, M.S. and Shear, M.D. (2015). Too Small for Radar to Detect, Rattles the White House, Jan 2015. https://www.nytimes.com/2015/01/27/us/white-house-drone.html (accessed 20 September 2021).

30 Wikipedia (2019). Gatwick Airport Drone Incident. December 2019. https://en.wikipedia.org/wiki/Gatwick_Airport_drone_incident (accessed 20 Septmebr 2021).

31 Nassi, B., Shamir, A., and Elovici, Y. (2019). Xerox day vulnerability. *IEEE Transactions on Information Forensics and Security* 14 (2): 415–430.

32 Guri, M., Zadov, B., and Elovici, Y. (2017). LED-it-GO: Leaking (a lot of) Data from Air-Gapped Computers via the (small) Hard Drive LED. *International Conference on Detection of Intrusions and Malware, and Vulnerability Assessment*, pp. 161–184. Springer.

33 Lindley, J. and Coulton, P. (2015). Game of drones. *Proceedings of the 2015 Annual Symposium on Computer-Human Interaction in Play*, pp. 613–618.

34 Toh, J., Hatib, M., Porzecanski, O., and Elovici, Y. (2017). Cyber security patrol: detecting fake and vulnerable WiFi-Enabled printers. *Proceedings of the Symposium on Applied Computing*, pp. 535–542.

35 Ronen, E., Shamir, A., Weingarten, A.-O., and O'Flynn, C. (2017). IoT goes nuclear: creating a ZigBee chain reaction. *2017 IEEE Symposium on Security and Privacy (SP)*, pp. 195–212. IEEE.

36 Angelov, P. (2012). *Sense and Avoid in UAS: Research and Applications*. Wiley.

37 Stevens, M.N., Coloe, B., and Atkins, E.M. (2015). Platform-independent geofencing for low altitude UAS operations. *15th AIAA Aviation Technology, Integration, and Operations Conference*, p. 3329.

38 Carroll, E.A. (2004). Miniature, unmanned aircraft with automatically deployed parachute. US Patent 6, 685, 140, 3 February 2004.

39 Taha, B. and Shoufan, A. (2019). Machine learning-based drone detection and classification: state-of-the-art in research. *IEEE Access* 7: 138669–138682.

40 Guvenc, I., Koohifar, F., Singh, S. et al. (2018). Detection, tracking, and interdiction for amateur drones. *IEEE Communications Magazine* 56 (4): 75–81.

41 Wyder, P.M., Chen, Y.-S., Lasrado, A.J. et al. (2019). Autonomous drone hunter operating by deep learning and all-onboard computations in GPS-denied environments. *PLoS ONE* 14 (11): e0225092.

42 Rothe, J., Strohmeier, M., and Montenegro, S. (2019). A concept for catching drones with a net carried by cooperative UAVs. *2019 IEEE International Symposium on Safety, Security, and Rescue Robotics (SSRR)*, pp. 126–132. IEEE.

43 Pleban, J.-S., Band, R., and Creutzburg, R. (2014). Hacking and securing the AR. Drone 2.0 Quadcopter: Investigations for improving the security of a toy. *Mobile Devices and Multimedia: Enabling Technologies, Algorithms, and Applications 2014*, Volume 9030, p. 90300L. International Society for Optics and Photonics.

44 Wesson, K. and Humphreys, T. (2013). Hacking drones. *Scientific American* 309 (5): 54–59.

45 Snead, J., Seibler, J.-M., and Inserra, D. (2018). *Establishing a Legal Framework for Counter-Drone Technologies*. Heritage Foundation.

46 Gray, C. (2002). Net launching tool apparatus. US Patent App. 09/814,527, 26 September 2002.

47 Armstrong, M.J., Hutchins, G.R., and Wachob, T.A. (2019). Interdiction and recovery for small unmanned aircraft systems. US Patent 10, 401, 129, 3 September 2019.

48 O'Malley, J. (2019). The no drone zone. *Engineering & Technology* 14 (2): 34–38.

49 Multerer, T., Ganis, A., Prechtel, U. et al. (2017). Low-cost jamming system against small drones using a 3D MIMO radar based tracking. *2017 European Radar Conference (EURAD)*, pp. 299–302. IEEE.

50 European Union Aviation Safety Agency (2020). Commission Delegated Regulations (EU) 2020/1058. https://eur-lex.europa.eu/legal-content/EN/TXT/PDF/?uri=CELEX:32020R1058\LY1\textbackslash&from=EN (accessed 20 September 2021).

51 Federal Aviation Administration (2021). UAS Remote Identification Overview. https://www.faa.gov/uas/getting_started/remote_id/ (accessed 20 September 2021).

52 ASTM International (2019). ASTM F3411 - 19: Standard Specification for Remote ID and Tracking. http://www.astm.org/cgi-bin/resolver.cgi?F3411-19 (accessed 20 September 2021).

53 Wi-Fi Alliance (2020). Wi-Fi Aware™ Specification Version 3.2.

54 FAA. Remote Identification of Unmanned Aircraft Systems. https://www.federalregister.gov/documents/2019/12/31/2019-28100/remote-identification-of-unmanned-aircraft-systems (accessed 20 September 2021).

55 ANSI/CTA (2019). ANSI/CTA-2063-A-Small Unmanned Aerial Systems Serial Numbers. https://standards.cta.tech/apps/group_public/project/details.php?project_id=587 (accessed 20 September 2021).

56 Li, L., Mathias, A.G., and Juang, B.-H. (2014). Provisioning an Embedded Subscriber Identity Module. US Patent 8, 843, 179, 23 September 2014.

57 Pannell, I. and eSIM Whitepaper (2018). The what and How of Remote SIM Provisioning. https://www.gsma.com/esim/wp-content/uploads/2018/06/eSIM-Whitepaper-v4.11.pdf (accessed 21 September 2021).

58 Luo, Y.B., Wang, B.S., and Cai, G.L. (2015). Analysis of port hopping for proactive cyber defense. *International Journal of Security and its Applications* 9 (2): 123–134.

59 Forsby, F. (2017). Digital Certificates for the Internet of Things.

60 Sehgal, R. and Sanjib, S. (2018). eSIM - Gateway to Global Connectivity.

61 Yeun, C.Y. (2000). Design, analysis and applications of cryptographic techniques. PhD thesis. University of London.

62 Dalal, N., Shah, J., Hisaria, K., and Jinwala, D. (2010). A comparative analysis of tools for verification of security protocols. *International Journal of Communications, Network and System Sciences* 3 (10): 779.

63 Blanchet, B., Smyth, B., Cheval, V., and Sylvestre, M. (2018). ProVerif 2.00: automatic cryptographic protocol verifier, user manual and tutorial.

64 Al Hamadi, H.M.N., Yeun, C.Y., Zemerly, M.J. et al. (2012). Verifying mutual authentication for the DLK protocol using ProVerif tool. *International Journal for Information Security Research* 2 (1): 256–265.

5

Collaborative Intrusion Detection in the Era of IoT: Recent Advances and Challenges

Wenjuan Li[1] and Weizhi Meng[2]

[1] *Department of Electronic and Information Engineering, The Hong Kong Polytechnic University, Hong Kong, China*
[2] *DTU Compute, Technical University of Denmark, Kgs. Lyngby, 2800, Denmark*

5.1   Introduction

Internet of Things (IoT) is currently transferring the conventional networks by allowing various devices to connect with each other, and the continued growth of IoT networks can benefit many organizations such as enterprises, consumers, and governments. The Gartner Report also indicates that the IoT market may grow to 5.8 billion endpoints in 2020, resulting in a 21% increase from 2019 [1]. An industry IoT Report estimates that the IoT market is on pace to grow to over 3 trillion annually by 2026, and that there will be more than 64 billion IoT devices by 2025 [2].

IoT devices like smart sensors can be remotely monitored and controlled, which provide a basis for building a smart city infrastructure. Through involving the information and communication technology (ICT) and IoT, a smart city seeks to improve the operational efficiency and quality of urban services to save more resources [3]. For instance, it is possible for IoT networks to help gain completely new insights on how to manage traffic by designing an intelligent traffic controller.

However, according to a survey with over 100 IoT leaders, IoT security could rank top two of the major concerns [4]. Up to 29% of the respondents believed that some form of "trust" should be established in IoT networks, and 15% leaders had a concern that IoT architectures rely heavily on centralized servers to collect and store data. Hence, there is a need to deploy appropriate security mechanisms to protect IoT environments.

Due to the distributed architecture of IoT networks, *collaborative intrusion detection networks* – CIDNs (or called *distributed intrusion detection systems* – DIDSs)

Security and Privacy in the Internet of Things: Architectures, Techniques, and Applications,
First Edition. Edited by Ali Ismail Awad and Jemal Abawajy.

are becoming a promising solution, which can deploy various detectors and allow them to exchange required information in order to enhance the detection performance [5, 6]. More specifically, an intrusion detection system (IDS) can be either host-based or network-based, based on the data source. According to the detection methodologies, an IDS can be roughly classified as signature-based detection and anomaly-based detection [7–9]. The former identifies attacks by comparing the incoming events with the patterns in the rule database. A pattern (or rule) is extracted from known attacks. On the other hand, the latter detects anomalies by modeling the normal behavior of a target system or network and then by monitoring any great deviations.

Currently, CIDNs/DIDSs have been used to address security issues in various areas, such as IoT, Li et al. [7]; healthcare, Meng et al. [10]; vehicles, Guo et al. [11]; and smart grid, Hu et al. [5]. However, such distributed detection framework is also a target by cyber-criminals and is particularly vulnerable to insider attacks where an attacker has the right to access the resources within a network/a system.

To address this challenge, constructing proper trust mechanisms is a necessary and effective way to help measure the trustworthiness of an inside detector. For instance, challenge-based mechanism is a promising solution to identify malicious nodes by sending a kind of message called *challenge* [12, 13]. In the era of IoT, with more devices to be inter- or intraconnected, there is a demanding need to design a more robust and suitable collaborative detection system. Motivated by this, this chapter aims to survey the recent development of collaborative intrusion detection systems/networks (CIDNs/CIDSs) in different disciplines. We then highlight some advanced insider attacks that may bypass the existing trust mechanisms and present some promising solutions. We also discuss existing open challenges and future directions in this area.

Chapter Organization. The next section introduces the background of IDS and CIDN including the core workflow and major components. Then, the third section introduces the recent development of CIDNs/CIDSs from 2003 to 2021. Then, the fourth section presents advanced insider threats and discusses existing open challenges with insights on future directions.

5.2 Background

This section first introduces the background of an IDS with the typical workflow and then illustrates the key components in a CIDN.

5.2.1 Background on Intrusion Detection System

Intrusion detection seeks to monitor the events in a computer network and analyze them for any signs of possible intrusions, which are not in-line with the defined

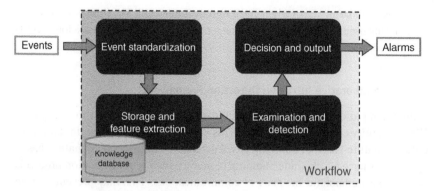

Figure 5.1 Core workflow of an IDS.

security policies or standard security practices [14]. As such, an IDS is the implementation that automates the intrusion detection process.

Figure 5.1 shows the main core workflow of an IDS, including event standardization, storage and feature extraction, examination and detection, and decision and output.

- *Event standardization.* Event here refers to either a local system event or a network event. To ensure that various events can be handled by an IDS, the first step is to standardize the format of all events. In a distributed environment, it is important to make sure that the events from different detectors can be managed.
- *Storage and feature extraction.* To analyze the incoming events, the second step is to extract predefined features from the events. This can ease the various detection approaches to perform an examination. For example, in the KDD dataset [15], there are nine basic features from individual transmission control protocol (TCP) connections and nine traffic features in a two-second time window.
- *Examination and detection.* The primary detection methodologies are signature-based and anomaly-based detection. Signature-based detection is the process of comparing signatures against observed events to identify possible threats. Anomaly-based detection is the process of comparing the observed events with predefined normal behavior to identify significant deviations. Signature-based approach is very effective at detecting known attacks but is mostly ineffective at detecting previously unknown attacks. By contrast, anomaly-based methods can be very effective at detecting previously unknown attacks, but may produce massive false alarms.
- *Decision and output.* In terms of specific detection methods, an IDS can make a decision accordingly. Taking anomaly-based detection as an example, machine learning can be involved in decision-making process. Nevertheless, the key point

here is to make an alarm when any anomaly is detected. The final output is a set of alarms used to notify security managers about the potential malicious events in the computer network.

5.2.2 Collaborative Intrusion Detection Framework

With the rapid evolution of adversarial techniques, it is difficult for a separate IDS to identify complex attacks [16]. This is also motivated by the limitations of either signature-based or anomaly-based detection. For instance, signature-based approaches have a little understanding of many network or application protocols and cannot understand the state of communications. This may prevent it from detecting attacks that comprise multiple events, like distributed denial-of-service (DDoS) attacks [17]. For anomaly-based detection, it has the difficulty in building normal profiles if the computing activity is complex. This will result in many false alarms especially in a diverse or dynamic environment.

To mitigate this challenge, a collaborative/distributed detection system, such as CIDN, CIDS, and DIDS, is developed to enhance the detection performance by allowing IDS nodes/detectors to communicate with each other and exchange important information. Based on the previous works of [6, 12, 16], Figure 5.2 depicts the typical framework of CIDNs with its major components including

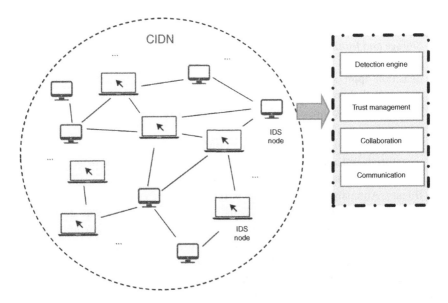

Figure 5.2 CIDN framework with major components.

detection engine, trust management component, collaboration component, and communication component.

- *Detection engine.* Similar to a single IDS, detection engine here can employ both signature-based and anomaly-based detection to help examine incoming events. In addition, the engine can also use stateful protocol analysis [14] by comparing predetermined profiles of generally accepted definitions of benign protocol activity, which is similar to a whitelist approach. Differently, the engine in a CIDN can receive the information from other nodes like alarms for better detection accuracy.
- **Trust management.** Insider threats are a big challenge to CIDNs/DIDSs; hence, most existing collaborative detection systems implement a trust management component to help build the trust relationship among various nodes and identify malicious nodes. For instance, challenge-based mechanism is a kind of trust management that aims to measure the reputation of nodes by sending challenges [12, 13].
- *Collaboration.* This component is used to help coordinate the information exchange among different nodes. As an example, it can deliver an alarm to its collaborated nodes for diagnosis if there is not enough information to make a decision. It is also an important component that collects the feedback from other nodes to measure the reputation.
- *Communication.* This component is mainly responsible for building physical connections with different nodes, e.g. P2P link.

For the interaction, each node can select its own collaborated nodes based on its own rules and experience [12, 13]. The IDS nodes can be associated if they have a collaborative relationship. Each node can maintain a list of partner nodes, which can be updated in a periodical manner. If a node needs to join the CIDN, it has to get its unique identity via a trusted third party (e.g. certificate authority). Typically, two types of messages can be used in a CIDN/DIDS.

- A *normal request* used for alarm aggregation can be sent during the interaction, which is an important feature of collaborative detection in improving the detection accuracy. The aggregation process usually only considers the feedback from highly trusted nodes. Therefore, an IDS node has to send back alarm ranking information as their feedback.
- A *measurement message* is used for evaluating the reputation of a node, which can be decided according to the concrete trust management models. Taking challenge-based trust mechanism as an example, a challenge can contain a set of IDS alarms, and be sent to the tested nodes for labeling alarm severity [12]. The trustworthiness of a node can be measured by identifying the difference between the received feedback and the expected answer.

5.3 Recent Development of Collaborative Intrusion Detection

With the rapid development of computer networks, many existing information infrastructures integrate with distributed systems with across multiple platforms, like system and myriad application level. In the era of IoT, the open and connected architecture should touch many untrusted clients over untrusted networks. To handle this issue, there is a significant need for involving a distributed or collaborative detection system or network.

Table 5.1 shows a summary of major studies on collaborative intrusion detection published from 2003 to 2019 in known journals and conferences.

In 2003, Wu et al. [16] introduced an early model called Collaborative Intrusion Detection System (CIDS), which deploys multiple specialized detectors in different layers of a computer system. They also implemented a management component to help collect alarms from different detectors and synthesize a global/aggregate alarm. The aggregate alarm is believed to be more accurate than an elementary alarm, i.e. it can decrease false alarms and increase the probability of finding missed alarms. The proposed CIDS focuses more on host-based detection, including three elementary detectors: *Snort* – a network-level detector, *Libsafe* that is an application-level detector, and *Sysmon* that is a kernel-level detector. They evaluated the system under a real-world web application and three classes of attacks. It was found the proposed system could reduce the incidence of missing alarms and false alarms with negligible impact on the performance.

Motivated by Wu et al. [16], more research on collaborative intrusion detection has appeared. Yeom and Park [18] introduced a bioinspired CIDS based on mobile agents, which coupled with a human immune system to handle the large and complex network environments. The used immune system contains a complex network of cells that are responsible for defending the body against external attacks. A good immune system should have a capability to distinguish own molecules (e.g. self) and foreign molecules (e.g. nonself). Their study proved the feasibility, but there is a lack of a prototype.

Meanwhile, Zhou et al. [19] proposed some improvements over the early developed CIDS, in the aspects of data routing, load balancing, scalability, and central points of failure. They particularly introduced a fully distributed CIDS based on a peer-to-peer architecture. Each peer submits its own suspicious IP addresses list, which is generated from its own network scope, to a collaborative node (or a system) that handles all the submitted information. A notification would be sent to the security manager if a match is found among different IP addresses. They implemented the system in a peer-to-peer architecture called OPeN, and it was found that their system could outperform the centralized system by reducing the detection latency.

Table 5.1 Development summary of collaborative intrusion detection.

Publication	Description
Wu et al. [16]	Introduction of an early model of CIDS
Yeom and Park [18]	A bioinspired CIDS based on immune system
Zhou et al. [19]	A fully distributed peer-to-peer CIDS
Beyah et al. [20]	A CIDS for wireless local area networks
Zhou et al. [21]	*LarSID*, a scalable and decentralized CIDS
Chaki and Chaki [22]	A cluster-based CIDS in mobile ad hoc network
Zhou et al. [23]	A load-balancing CIDS
Marchang and Datta [24]	A CIDS for mobile ad hoc networks
Fung et al. [25]	Host-based IDS (HIDS) with Bayesian trust management model
Fung et al. [26]	A CIDS with Bayesian feedback aggregation
Zargar and Joshi [27]	A CIDS within an autonomous system
Fung et al. [28]	A CIDS with Dirichlet-based trust management
Steven [29]	A CIDS against statistic-poisoning attacks
Gu et al. [30]	A CIDS for multichannel mesh networks
Pérez et al. [31]	A reputation-based CIDN
Li et al. [32]	A CIDN with detection sensitivity
Morais and Cavalli [33]	A CIDS for Wireless Mesh Networks (WMNs)
Kumar and Chilamkurti [34]	T-CLAIDS, a CIDS for VANETs
Cordero et al. [35]	A CIDS with ensemble learning
Vasilomanolakis et al. [36]	A probe-response attack on CIDSs
Liu et al. [37]	A CIDS for smart grid
Andreolini et al. [38]	A stateful CIDS
Chen and Yu [39]	A CIDS for SDNs
Fung and Zhu [40]	FACID, a CIDN with feedback aggregation
Li et al. [41]	An intrusion sensitivity-based CIDN
Patel et al. [42]	A CIDS/CIPS in smart grids
Tug et al. [43]	Blockchain-based CIDNs
Cordero et al. [44]	Sphinx, an evidence-based CIDS
Hong and Liu [45]	A CIDS for IEDs
Arshad et al. [46]	COLIDE, a CIDS for IoT networks
Hu et al. [5]	A CIDS for Multimicrogrid
Meng et al. [47]	A blockchain-based CIDS for IoMT

Beyah et al. [20] gave a trial of applying a CIDS for wireless local area networks (WLANs). They believed that CIDS could protect the wireless network from unauthorized users. Hu et al. [48] focused on CIDS and investigated how to make a trade-off between cost and monitoring coverage, with the purpose of deciding the positions and processing rates of each sensor. They then built fuzzy expected value optimization models with a hybrid intelligent algorithm. Based on their method, it could realize using a small number of low-speed sensors to maintain a high monitoring coverage.

Zhou et al. [21] aimed to address the question on how to detect large-scale coordinated attacks such as scans, worms, and denial-of-service attacks. They then proposed *LarSID*, a scalable decentralized large-scale intrusion detection framework, which can provide a service for defeating various attacks by sharing potential evidence of intrusions between nodes via a distributed hash table (DHT) architecture. They evaluated their system using a large and real-world IDS dataset (DShield Dataset), regarding how to optimize the trade-off between detection accuracy and reaction time. Their results presented a significant reduction in detection latency as compared with a centralized architecture. Chaki and Chaki [22] proposed a cluster-based algorithm for CIDSs in mobile ad hoc network.

Zhou et al. [23] identified that a decentralized CIDS architecture may generate load hot spots among participant IDSs, when all the suspicious traffic comes from the same source during a very short frame, e.g. during a worm outbreak, for example when an infected victim machine starts infecting other nodes by massively scanning the whole network and locating a specific software vulnerability. This node then becomes a load "hot spot" in the CIDS, which may cause delays in alarm correlation or even information loss. In addition, an attacker can exploit this vulnerability by launching many kinds of scan attacks to multiple nodes with the aim to overload the processing capability of others and finally harm the whole CIDS environment. For this issue, they introduced a load balancing scheme for CIDSs, which evenly distributes workload during worm outbreaks, through distributing suspicious evidence by indexing according to each possible alert pattern instance. In the evaluation, their approach could reach a high detection rate (more than 90%) and negligible false positive (FP) rate (less than 0.12%) during worm outbreaks.

Marchang and Datta [24] applied CIDS to mobile ad hoc networks, which could help detect two kinds of threats: (i) Detection of malicious nodes in a neighborhood of nodes in which each pair of nodes in the neighborhood are within radio range of each other. (ii) Detection of malicious nodes in a neighborhood of nodes, in which each pair of nodes may not be in radio range of each other, but there exists a node among them that has all the other nodes in its one-hop vicinity. They particularly designed two algorithms: one is called Algorithm for Detection in a CLIque (ADCLI), and the other one is called Algorithm for Detection in a

CLUster (ADCLU). Both algorithms used a message-passing mechanism between the group of nodes, which enable each of the nodes to determine those nodes in the group that are suspected to be malicious. Their results showed that their algorithms could perform well when there is a packet loss rate less than 5%.

Fung et al. [49] noticed the insider threat for a CIDS and proposed a trust management model for distributed HIDS collaboration. Their trust management model allows a node sending a message called challenge to evaluate the trustworthiness of other nodes. The framework gave incentives to motivate collaboration. For instance, nodes that are consulted would only reply to a number of requests in a certain period of time due to the limited bandwidth and computational resources. In this case, only highly trusted nodes have higher priority of receiving required information or data whenever needed. This aims to encourage insider nodes to maintain their reputation.

Marchetti et al. [50] proposed a CIDS framework for malware analysis and alert dissemination, including three major layers: the sensor layer, the local aggregation layer, and the collaboration layer. The proposed architecture is based on a decentralized, peer-to-peer, and sensor-agnostic design that addresses dependability and load unbalance issues affecting existing systems based on centralized and hierarchical schemes. The experiment showed that their approach could preserve the same load distribution even under attacks.

Fung et al. [25] extended their previous work and designed a robust Bayesian trust management model for distributed HIDS collaboration. Their method adopts the Dirichlet family of probability density functions for trust management, with the purpose of estimating the possible future behavior of HIDSs based on the collected past history. As it can track the uncertainty in estimating the trustworthiness of HIDS nodes, the detection accuracy can be improved. Their experimental results showed that the system could provide a significant improvement in detecting attacks and is more scalable as compared with previous systems.

Zhou et al. [51] focused on the alert correlation issue in a CIDS, that is, how to improve the scalability of alert correlation while still maintaining the expressiveness of the patterns that can be found. They then introduced a correlate-and-filter algorithm to analyze multidimensional alerts in a CIDS, based on a lattice of possible patterns that correspond to specific categories of attacks. They also gave a hierarchical two-stage scheme to distribute the computation in the CIDS, by analyzing raw alerts locally at each IDS before forwarding the results. They performed an evaluation in a large-scale deployment on PlanetLab and found that their fully decentralized P2P CIDS is significantly more efficient than a centralized CIDS.

Fung et al. [26] designed a collaboration framework for CIDNs by applying a Bayesian approach for feedback aggregation, in order to minimize the combined costs of missed detection and false alarm. Their results demonstrated an improvement in the true positive detection rate and a reduction in the average cost than similar models. Zhu et al. [52] introduced a sequential hypothesis testing method for feedback aggregation in CIDNs. Their algorithm could help reduce the communication overhead until a predefined FP and true positive goal is reached. The analytical model can effectively estimate the number of acquaintances needed for an IDS node to reach the goals, which is very helpful in deciding how to build an IDS partner list.

Fung et al. [53] also targeted the same issue on how to select and maintain a list of partner nodes. In this system, each HIDS evaluates both the FP rate and false negative (FN) rate of its neighboring HIDSes' opinions about intrusions using Bayesian learning and aggregates their opinions about intrusions using a Bayesian decision model. Zargar and Joshi [27] focused on the detection of DDoS attacks and introduced a collaborative approach to distribute the sampling, detection, and response responsibilities among all the routers within an autonomous system (AS). Their method aimed to coordinate all the routers in the network to eliminate redundant sampling, detection, and response tasks without exploiting any specific communication protocol.

Grunewald et al. [54] designed a Distributed, Collaborative, and Data-driven Intrusion Detection and Prevention system (DCDIDP) by leveraging the resources in the cloud. It can collaborate with other peers in a distributed manner and respond to attacks at different architectural levels. Fung et al. [28] provided a Dirichlet-based trust management to help evaluate the level of trust among different nodes according to their mutual experience. They also implemented an acquaintance management algorithm to handle the partner nodes. Steven [29] identified that one big threat for CIDSs is statistic-poisoning attacks, where intruders might obfuscate attacks on the CIDS statistics and decision. To mitigate this issue, an approach was proposed for computing CIDS statistics in a more robust manner. It uses contributor-level aggregation and preferential voting in which voters rank the candidates and the voting algorithm computes a set of winners. In addition, a single transferable vote method is used for computing the CIDS statistics. The results on a subset of DShield dataset showed that the voting-based approach could be more resistant against attacks.

Gu et al. [30] studied on the detection issue in multichannel mesh networks in which adversarial actions may not be observed as some ad hoc nodes communicate on different channels. To investigate the optimal channel choice, they introduced a collaborative TRaffic-Aware intrusion Monitoring (TRAM) scheme in a multi-channel mesh network to select the monitoring channels and coordinate transmission and monitoring with each other. They first model the problem of choosing monitoring channels as the TRAM problem and proved that it is NP-hard. Then they formulated the TRAM problem as a mathematical programming problem with several heuristic solutions by allowing mesh routers to choose channels based on their local information.

Pérez et al. [31] designed a reputation-based CIDN, which is a partially decentralized system to detect distributed attacks both in a local environment and in a large-scale network. The information has to be previously assessed by a Wise Committee (WC) before being shared with other nodes. Thus, no node can publish alerts without the prior consent of its WC, thereby avoiding the spread of alerts generated by malicious or compromised IDSs. It can also improve the detection coverage by dropping false or bogus alarms. Fung and Boutaba [12] combined their previous work on CIDNs and provided an evaluation in the aspects of trust management, collaborative intrusion decision, resource management, and collaborators selection. Li et al. [32] then proposed a concept of detection sensitivity for CIDNs, which measures the capability of detecting different types of intrusions. In the evaluation, they showed that intrusion sensitivity can help enhance the accuracy of detecting malicious nodes.

Li et al. [55] further extended their former work and introduced a machine learning-based approach to help assign the value of intrusion sensitivity using expert knowledge. They then designed a trust management model that allows each IDS to evaluate the trustworthiness of others by considering their detection sensitivities. Morais and Cavalli [33] focused on a Wireless Mesh Network (WMN) and provided a distributed CIDS architecture for detecting insider attacks at real time. It has several major components: a *Routing Protocol Analyzer (RPA)* to analyze the collected routing traffic and generate respective routing events; a *Distributed Intrusion Detection Engine (DIDE)* that handles the Routing Events by applying Routing Constraints and calculating related Misbehaving Metrics; and a *Cooperative Consensus Mechanism (CCM)* to check the Misbehaving Metrics. Kumar and Chilamkurti [34] introduced a Trust aware Collaborative Learning Automata-based Intrusion Detection System (T-CLAIDS) for vehicular ad hoc networks (VANETs). They used a Markov Chain Model (MCM) to represent the states and relevant transitions in the network. A classifier is designed for identifying any malicious activity in the network and is tuned based on a parameter called Collaborative Trust Index (CTI) to measure the trustworthiness of a node.

Cordero et al. [35] identified that the collaboration by exchanging suspicious alarms among all interconnected nodes in a CIDS may not scale with the size of the IT infrastructure, and there is a need to trade off detection performance and communication overhead. They particularly found that the ensemble learning can be applied as a distributed machine learning method for a CIDS. Each node can share data with its communities so that subsets of the entire dataset are created. Their results indicated that the community-based CIDS can perform similarly to centralized systems even though less information is distributed while less overhead is involved. Vasilomanolakis et al. [36] introduced a probe-response attack on CIDSs and then presented a simple but effective metric to detect such kind of attack by utilizing the ratio of generated alerts in relationship to the number of actively reporting sensors.

Liu et al. [37] targeted on the smart meters in smart grids and proposed a CIDS to defend against false data injection attack by building the legitimate reading and writing operations. Andreolini et al. [38] first introduced a mobile evasion attack in which an intruder exploits network mobility to avoid detection even by the most advanced stateful NIDS systems. They then presented a CIDS solution that allows sharing of internal state information among multiple NIDSs deployed in different networks or network segments.

Li and Meng [56] measured the impact of intrusion sensitivity regarding the detection of pollution attacks in a CIDN. The experimental results demonstrate that the proposed approach is more effective in reducing the reputation of suspicious nodes under pollution attacks and maintaining the accuracy of alarm aggregation as compared to similar approaches. Chen and Yu [39] focused on the DDoS attacks on Software-Defined Networks and proposed a distributed CIDS that can be deployed as a modified artificial neural network distributed over the entire substrate of SDN. It disperses its computation power over the network that requires each participating switch to perform like a neuron. Chen and Yu [57] later introduced a collaborative intrusion prevention architecture. Fung and Zhu [40] presented FACID, a CIDN that leverages data analytical models and hypothesis testing methods for efficient, distributed, and sequential feedback aggregation.

Li et al. [41] extended the previous work and developed an intrusion sensitivity-based trust management model that allows each IDS to evaluate the reputation of others by considering their detection sensitivities. They also developed a supervised approach to automatically assign the values of intrusion sensitivity based on expert knowledge. In the evaluation, they evaluated their model in a real wireless sensor network and found that it can enhance the detection accuracy of malicious nodes and achieve better performance as compared with similar models. Patel et al. [42] focused on smart grids and introduced a smart collaborative advanced intrusion prevention system including a fuzzy risk analyzer and an independent and ontology knowledge-based inference engine module. They showed that a rich ontology knowledge base with fuzzy logic analysis can detect and prevent intrusions more efficiently.

Zhang and Zhu [58] focused on a VANET and studied some collaborative detection algorithms using machine learning. In order to address the privacy concern during data exchange, they provided a privacy-preserving machine-learning-based collaborative intrusion detection system (PML-CIDS). They employed the alternating direction method of multipliers to a class of empirical risk minimization problems and trained a classifier to detect the intrusions in the VANETs. Tug et al. [43] introduced a generic framework called collaborative blockchained signature-based IDSs, which utilizes blockchain technology to incrementally build a trusted signature database for different nodes in a CIDS. Their evaluation results showed that blockchain technology could indeed help enhance the robustness and effectiveness of signature-based IDSs under adversarial scenarios.

Cordero et al. [44] aimed to identify insider attacks by building a trust mechanism that detects honest and dishonest nodes in terms of their reliability. They particularly focused on a group of coordinated dishonest nodes (known as coalition) that cooperate with each other to improve their trust while reducing the reputation of other nodes. To address this issue, they proposed Sphinx, an evidence-based trust mechanism that uses the sensing reliability of participating nodes to detect dishonesty. In the evaluation, they showed that Sphinx could detect single large coalitions as long as the majority of IDS nodes are honest. When there are multiple independent coalition, it also can detect dishonesty if the largest independent coalition is smaller than the number of honest nodes.

Zhou and Pezaros [59] introduced Bioinspired, collaborative intrusion detection for SDN (BIDS), an artificial immune system (AIS) based CIDS. The distributed AIS IDS is running on each switch in a network while an antibody fuser running on the controller orchestrates the training results from all the switches. After being trained locally, the controller will collect mature antibody sets trained and the intrusion set recorded by each switch. These sets then will be fused by the antibody fuser as one global mature antibody set and subsequently synced with all switches. Hong and Liu [45] focused on intelligent electronic devices (IEDs) and designed a CIDS. In particular, they used the specification-based detection algorithm to detect simultaneous attacks at the same time.

Vasilomanolakis and Mühlhäuser [60] focused on probe-response attacks on CIDSs in which an attacker has the ability to detect the network location of the monitors within a CIDS. Arshad et al. [46] introduced a CIDS framework (COLIDE) for IoT environments by leveraging collaboration among resource-constrained sensor and border nodes to detect attacks. The implemented system was realized by using the COLIDE framework with Contiki OS. Their evaluation demonstrates the efficiency in the aspects of energy efficiency and low-processing overhead in an IoT system. Hu et al. [5] presented a CIDS for multimicrogrid (MMG) using blockchains. It can provide the consistency and nonrepudiability of detection results in each microgrid during the distributed data transmission.

Meng et al. [61] found the burden of signature matching in IoT environments and proposed a blockchain-enabled single character frequency-based exclusive signature matching, which can build a verifiable database of malicious payloads via blockchains. Meng et al. [47] focused on the Internet of Medical Things (IoMT) and used blockchains to improve the effectiveness of Bayesian inference-based trust management to detect malicious nodes. Experimental results demonstrate that blockchain technology can help improve the detection efficiency of detecting malicious nodes with reasonable workload. Ma et al. [62] targeted on a type of multiple-mix-attack with a combination of three typical attacks – tamper attack, drop attack, and replay attack. Then they proposed Distributed Consensus-based Trust Model (DCONST), an approach that can evaluate the trustworthiness of IoT nodes by sharing certain information called cognition. They also designed three modes of DCONST: DCONST-Light, DCONST-Normal, and DCONST-Proactive, and the experimental results show that DCONST-Normal and DCONST-Proactive can further improve the detection rate by 5– 20% as compared to DCONST-Light.

Li et al. [63] introduced a generic framework aiming to enhance the security of CIDSs against advanced insider threats by deriving multilevel trust. Similarly, Li et al. [64] advocated that the combination of additional trust can enhance the robustness of CIDNs and introduced an enhanced trust management scheme by checking spatial correlation among nodes' behavior, regarding forwarding delay, packet dropping, and sending rate. Experimental results demonstrate that their approach could help enhance the robustness of challenge-based trust mechanism under IoT environments. Meng et al. [65] introduced a type of blockchain-based trust to help enhance the robustness of challenge-based CIDNs against advanced insider attacks. In the evaluation with both simulated and real network environments, their approach was found to be effective in defeating advanced insider attacks and enhancing the robustness of challenge-based CIDNs, as compared with the original scheme. Li et al. [66] found that labeled instances are quite limited for intrusion detection in real-world IoT scenarios, and designed DAS-CIDS, by applying disagreement-based semisupervised learning algorithm. The experimental results showed that as compared with traditional supervised classifiers, DAS-CIDS was more effective in detecting intrusions and reducing false alarms by automatically leveraging unlabeled data.

5.4 Open Challenges and Future Trend

5.4.1 Advanced Insider Threats

With the rapid development over almost two decades, collaborative intrusion detection has become more robust by involving trust management. However, existing trust mechanisms may be still vulnerable to some advanced attackers, e.g. APT attackers. We introduce some advanced insider attacks and relevant solutions as below.

5.4.1.1 Advanced Attacks
Meng et al. [67] introduced a kind of random poisoning attack on challenge-based CIDSs, where an attacker can behave maliciously in a random manner. The experimental results demonstrate that such an attack can enable a malicious node to

send untruthful information without decreasing its trust value at large. Li et al. [68, 69] proposed a kind of advanced collusion attack called *passive message fingerprint attack (PMFA)* on the challenge-based CIDSs, which can collect messages and identify normal requests in a passive way. In the evaluation, it is found that under such an attack, malicious nodes can send malicious responses to normal requests while maintaining their trust values.

Li et al. [70, 71] introduced a special On-Off attack on challenge-based CIDNs, shortly special on-off attack (SOOA), in which a SOOA node can keep responding normally to one node while acting abnormally to another node. In the evaluation, they explored the attack performance under simulated CIDN environments and found that such an attack can interfere with the effectiveness of trust computation for CIDN nodes. Meng et al. [72] designed a collusion attack called Bayesian Poisoning Attack, which enables a malicious node to model received messages using Bayesian inference and then craft a malicious response to those messages, which have higher aggregated appearance probability to be normal requests. In the evaluation under both simulated and real network environments, their results demonstrate that the malicious nodes under such attack can successfully craft and send untruthful feedback while maintaining their reputation.

5.4.1.2 Solutions

To mitigate these threats, how to build a more robust trust mechanism remains a challenge. Li et al. [73] focused on PMFA and designed Honey Challenge, an improved challenge mechanism for challenge-based CIDNs characterized by sending challenges in a similar way of sending normal requests, in such a way that malicious nodes cannot accurately identify the normal requests. Tug et al. [43] also targeted on PMFA and introduced a compact but efficient message verification approach by inserting a verification alarm into each normal request. In the evaluation, they investigated the attack performance under both simulated- and real-network environments. The results demonstrate that their approach can identify malicious nodes under PMFA and decrease their trust values in a quick manner.

Madsen et al. [74] evaluated the impact of intrusion sensitivity on detecting SOOA, which can highlight the detection capability of expert nodes on particular attacks. Their experimental results demonstrated that the use of intrusion sensitivity can help enhance the security of CIDNs under SOOA. Li et al. [7, 13] recently provided a blockchained challenge-based CIDN framework by combining CIDN with blockchains. Their evaluation showed that blockchain technology can help enhance the robustness of challenge-based CIDNs in the aspects of both trust management and alarm aggregation, i.e. detecting malicious inputs via blockchains.

Li and Kwok [6] further analyzed the effectiveness of PMFA and investigated whether an improved sending strategy can help detect malicious nodes. The study reveals that PMFA could still be valid under even an improved sending strategy, if malicious nodes can hold its reputation level by understanding the network context. This indicates that an effective solution should be designed according to a specific insider threat.

5.4.2 Open Challenges and Limitations

Collaborative intrusion detection has the capability of enhancing the detection performance against various attacks, and trust management schemes can help improve the robustness of CIDNs/DIDSs against insider threats. In the literature, there are still many open challenges remained unsolved.

- *Incentive mechanism.* The purpose of this mechanism is to motivate and reward sensors for behaving in a benign manner. For example, an incentive can be given to only those sensors who own high trust values, i.e. they can provide feedback on alarm aggregation process. However, a nonsuitable incentive mechanism may degrade the detection performance. Tuan [75] figured out that if a reputation system was not incentive-compatible, the more peers in the system, the less likely that anyone would report a malicious peer. Thus, it is an open challenge on how to design a proper incentive mechanism in practice.
- *Initial trust.* It is still a basic issue on how to assign an initial trust value to an IDS node. Pérez et al. [76] targeted on this issue and tried to solve two questions of cold-start and reputation bootstrapping. Cold-start is a common issue to any system when newcomers boot for the first time, while reputation bootstrapping especially affects highly distributed scenarios, where mobile entities travel across domains and collaborate with a number of them. In a CIDN, since historical data are not always available, a trust mechanism can choose either assigning random values or a probation period approach. As each approach has its own advantages and disadvantages, how to assign an initial trust value needs considering a practical scenario.
- *Impact of historical data.* With more historical data, the impact of such old data may decrease the detection sensitivity. Performance history can be used to describe how a sensor has performed based on historical data. Thus, a forgetting factor can be used to assign less weight on the historical data when performing trust evaluation, while it does not mean that a better trust relationship can be built by involving many more recent data. In practice, some noise data should be carefully filtered to ensure that effective data are applied for computing trust values.
- *Trust computation.* How to measure the reputation of nodes is an open challenge when designing a trust management scheme. For challenge-based mechanism, IDS alarms are used for such a purpose, while it is a good question to

investigate additional parameters, e.g. inside the alert data, to be used in trust computation [77]. In addition, it may consider other social attributes or network connection features for building a trust computation process.

- *Alarm exchange.* The main purpose of trust mechanism is to evaluate the reputation of other nodes and only highly trusted nodes can exchange alarms for aggregation. However, there are still some issues remained: (i) What is the impact of timeliness of the exchanged alarm data on the performance. (ii) What is the impact of the relevance of the exchanged alarms data on the performance, i.e. whether it is valid if an alarm is received from a highly trustworthy but irrelevant node.

- *Alarm aggregation.* The capability of aggregating alarms is an important advantage of collaborative detection, which can greatly enhance the detection performance over a single detector. To avoid the impact from malicious insider nodes, alarm aggregation only considers receiving the alarm information from highly trusted nodes. However, existing trust management models may be still vulnerable to advanced attacks, e.g. PMFA [6, 69], SOOA [70, 74]. That means advanced intruders can maintain their reputation and keep making an impact on the aggregation process. This is an open challenge for most current trust mechanisms.

- *Communication overhead.* In a CIDN/DIDS, exchanging information and data would cause overhead. For instance, challenge-based CIDNs exchange test messages like alarms and also use challenges to evaluate the trustworthiness of other nodes. The overhead may be also caused by detector reconfiguration. To reduce the overhead, trust computation can use the knowledge gained from past interactions [78], but this cannot ensure generating a robust trust value. How to balance the overhead and the accuracy of trust computation still remains an open challenge.

- *Attack probability.* An advanced attacker may choose a strategy to launch an attack, that is, an attack may be launched in a probability rather than all the time. How to include such an uncertainty into a trust model is an open challenge. To create a certainty value, Habib et al. [79] described an approach that could consider generating a trust value by considering the volume of data, but when there are no enough data, the certainty value would be low. Hence, it is a good research direction in this area.

5.4.3 Future Trend

With the rapid development of IoT networks, security has become a major concern. Collaborative intrusion detection is a necessary solution that allows distributed detectors to exchange required information. Below are some positive directions for designing a CIDN/DIDS in the near future.

- *Cyber physical system.* With the increasing scale of current networks, IoT would become a critical part even in some security-sensitive environments [80, 81]. Cyber physical system (CPS) refers to those computers that can interact directly with the physical world. The most common CPSs in our every day are modern cars in which computers can control not only the engine but also the braking and the vehicle stability. These systems are very complex, especially when several CPSs need to be combined. In such a case, a collaborative detection system is more beneficial to enhancing the detection of any malicious events [82]. This would become a research focus with the understanding on how important a CPS could be in current critical infrastructures.

- *Blockchain-based solution.* Due to the big success of Bitcoin, blockchain technology has received much attention from both industry and academia. At present, blockchain has the potential to become a bedrock of the worldwide record-keeping systems. The merit of bitcoin, as a kind of electronic cash, is that it could be sent to peer-to-peer without the need for a central authority like banks to operate and maintain the ledger. For collaborative intrusion detection, blockchain technology provides a solution for building a trust relationship among various (unknown) entities without the need of a trusted third party. Currently, there have been many relevant studies on blockchain-based CIDNs/DIDSs in the literature, like [5, 7, 83]. Because it is hard to find a (highly) trusted third party in practice, blockchain-based solution would become much more popular for designing future CIDNs.

- *Smartphone involvement.* In current IoT environments, smartphone is the most commonly used IoT devices for common people. They may use the smart devices everywhere, and everyone carries a smartphone all the time. Smartphone usage statistics indicate that a person spends two hours and 51 minutes per day on their mobile device on average. In particular, 22% of them may check the phones every few minutes, and 51% of users check the phones a few times every hour [84]. In future, there could be more research focusing on smartphone-based IoT networks like medical smartphone network (MSN) [85]. As smartphones provide a computing platform, many trust management models and trust computation schemes can be designed and applied.

IoT security threats have become more severe than ever; there is a demanding need for a strong, robust, and accurate CIDN/CIDS/DIDS in practice. Collaborative intrusion detection has made an impact in various areas such as healthcare [10, 85], vehicular [86], smart grid [87, 88], and so on. On the other hand, many new techniques can also complement the detection performance of collaborative detection, such as SDN [80], blockchain [5, 7], artificial intelligence [89], and edge computing [90]. Therefore, the design of a practical CIDN/DIDS should consider all these aspects and make a balance between performance and overhead.

5.5 Conclusion

In the era of IoT, more devices are connected, resulting in a more distributed network environment. To protect such networks, designing a CIDN/DIDS is an important solution. This chapter reviews the development of collaborative intrusion detection from 2003 to 2021, and identifies advanced insider attacks with relevant solutions. We also discuss open challenges and provide insights on future directions in this area. With the increasing IoT devices, there is a need for designing a suitable CIDN/CIDS/DIDS by balancing both performance and overhead in practice.

References

1 Goasduff, L. (2019). Gartner Says 5.8 Billion Enterprise and Automotive IoT Endpoints Will Be in Use in 2020. *Gartner Report*.

2 Newman, P. (2019). IoT report: how internet of things technology growth is reaching mainstream companies and consumers. *Business Insider Intelligence*.

3 Du, R., Santi, P., Xiao, M. et al. (2019). The sensable city: a survey on the deployment and management for smart city monitoring. *IEEE Communication Surveys and Tutorials* 21 (2): 1533–1560.

4 SCOOP (2019). IoT world 2019: the issues, plans and actions of IoT executives. https://www.i-scoop.eu/internet-of-things-guide/iot-world-2019-report/ (accessed 23 September 2021).

5 Hu, B., Zhou, C., Tian, Y.-C. et al. (2019). A collaborative intrusion detection approach using blockchain for multimicrogrid systems. *IEEE Transactions on Systems, Man, and Cybernetics: Systems* 49 (8): 1720–1730.

6 Li, W. and Kwok, L.F. (2019). Challenge-based collaborative intrusion detection networks under passive message fingerprint attack: a further analysis. *Journal of Information Security and Applications* 47: 1–7.

7 Li, W., Tug, S., Meng, W., and Wang, Y. (2019). Designing collaborative blockchained signature-based intrusion detection in IoT environments. *Future Generation Computer Systems* 96: 481–489.

8 Song, H. and Lockwood, J.W. (2005). Multi-pattern signature matching for hardware network intrusion detection systems. *Proceedings of the Global Telecommunications Conference, 2005. GLOBECOM '05*, St. Louis, Missouri, USA (28 November - 2 December 2005), p. 5. IEEE.

9 Engly, A.H., Larsen, A.R., and Meng, W. (2020). Evaluation of anomaly-based intrusion detection with combined imbalance correction and feature selection. In: *Network and System Security - 14th International Conference, NSS 2020, November 25–27, 2020, Proceedings, Lecture Notes in Computer Science,*

vol. 12570 (ed. M. Kutylowski, J. Zhang, and C. Chen), 277–291. Melbourne, VIC, Australia: Springer.

10 Meng, W., Li, W., Xiang, Y., and Choo, K.K.R. (2017). A Bayesian inference-based detection mechanism to defend medical smartphone networks against insider attacks. *Journal of Network and Computer Applications* 78: 162–169.

11 Guo, P., Kim, H., Guan, L. et al. (2017). VCIDS: Collaborative intrusion detection of sensor and actuator attacks on connected vehicles. In: *Security and Privacy in Communication Networks - 13th International Conference, SecureComm 2017, Niagara Falls, ON, Canada, October 22–25, 2017, Proceedings*, (eds. Xiaodong Lin, Ali A. Ghorbani, Kui Ren, Sencun Zhu, Aiqing Zhang), *Lecture Notes of the Institute for Computer Sciences, Social Informatics and Telecommunications Engineering*, vol. 238, 377–396. Springer.

12 Fung, C.J. and Boutaba, R. (2013). Design and management of collaborative intrusion detection networks. *2013 IFIP/IEEE International Symposium on Integrated Network Management(IM)*, Ghent, Belgium (27–31 May 2013), pp. 955–961. IEEE.

13 Li, W., Wang, Y., Li, J., and Au, M.H. (2019). Towards blockchained challenge-based collaborative intrusion detection. *Applied Cryptography and Network Security Workshops - ACNS 2019 Satellite Workshops, SiMLA, Cloud S&P, AIBlock, and AIoTS, Bogota, Colombia, June 5–7, 2019, Proceedings, Lecture Notes in Computer Science*, vol. 11605, pp. 122–139. Springer.

14 Scarfone, K. and Mell, P. (2007). *Special Publication 800-94: Guide to Intrusion Detection and Prevention Systems (IDPs)*. National Institute of Standards and Technology (NIST).

15 Meng, Y. (2011). The practice on using machine learning for network anomaly intrusion detection. *The 2011 International Conference on Machine Learning and Cybernetics (ICMLC)*.

16 Wu, Y.S., Foo, B., Mei, Y., and Bagchi, S. (2003). Collaborative intrusion detection system (CIDS): a framework for accurate and efficient IDS. *19th Annual Computer Security Applications Conference (ACSAC 2003)*, Las Vegas, NV, USA (8–12 December 2003), pp. 234–244. IEEE Computer Society.

17 Wang, M., Lu, Y., and Qin, J. (2020). A dynamic MLP-based DDoS attack detection method using feature selection and feedback. *Computers & Security* 88. 1–14.

18 Yeom, K.-W. and Park, J.-H. (2005). An immune system inspired approach of collaborative intrusion detection system using mobile agents in wireless ad hoc networks. In: *Computational Intelligence and Security, International Conference, CIS 2005, December 15-19, 2005, Proceedings, Part II*, (eds. Yue Hao, Jiming Liu, Yu-Ping Wang, Yiu-ming Cheung et al.), *Lecture Notes in Computer Science*, vol. 3802, 204–211. Xi'an, China: Springer.

19 Zhou, C.V., Karunasekera, S., and Leckie, C. (2005). A peer-to-peer collaborative intrusion detection system. *The 13th IEEE International Conference on Networks Jointly Held with the 2005, IEEE 7th Malaysia International Conference on Communication.*

20 Beyah, R.A., Corbett, C.L., and Copeland, J.A. (2006). The case for collaborative distributed wireless intrusion detection systems. *2006 IEEE International Conference on Granular Computing, GrC 2006, Atlanta, Georgia, USA (10–12 May 2006)*, pp. 782–787. IEEE.

21 Zhou, C.V., Karunasekera, S., and Leckie, C. (2007). Evaluation of a decentralized architecture for large scale collaborative intrusion detection. *Integrated Network Management, IM 2007. 10th IFIP/IEEE International Symposium on Integrated Network Management*, Munich, Germany (21–25 May 2007), pp. 80–89. IEEE.

22 Chaki, R. and Chaki, N. (2007). IDSX: A cluster based collaborative intrusion detection algorithm for mobile ad-hoc network. *6th International Conference on Computer Information Systems and Industrial Management Applications, CISIM 2007*, Elk, Poland (28–30 June 2007), pp. 179–184. IEEE Computer Society.

23 Zhou, C.V., Karunasekera, S., and Leckie, C. (2008). Relieving hot spots in collaborative intrusion detection systems during worm outbreaks. *IEEE/IFIP Network Operations and Management Symposium: Pervasive Management for Ubioquitous Networks and Services, NOMS 2008* (7–11 April 2008), Salvador, Bahia, Brazil, pp. 49–56. IEEE.

24 Marchang, N. and Datta, R. (2008). Collaborative techniques for intrusion detection in mobile ad-hoc networks. *Ad Hoc Networks* 6 (4): 508–523.

25 Fung, C.J., Zhang, J., Aib, I., and Boutaba, R. (2009). Robust and scalable trust management for collaborative intrusion detection. *Integrated Network Management, IM 2009. 11th IFIP/IEEE International Symposium on Integrated Network Management*, Hofstra University, Long Island, NY, USA (1–5 June 2009), pp. 33–40. IEEE.

26 Fung, C.J., Zhu, Q., Boutaba, R., and Basar, T. (2010). Bayesian decision aggregation in collaborative intrusion detection networks. *IEEE/IFIP Network Operations and Management Symposium, NOMS 2010*, Osaka, Japan (19–23 April 2010), pp. 349–356. IEEE.

27 Zargar, S.T. and Joshi, J.B.D. (2010). A collaborative approach to facilitate intrusion detection and response against DDoS attacks. *The 6th International Conference on Collaborative Computing: Networking, Applications and Worksharing, CollaborateCom 2010*, Chicago, IL, USA (9–12 October 2010), pp. 1–8. ICST / IEEE.

28 Fung, C.J., Zhang, J., Aib, I., and Boutaba, R. (2011). Dirichlet-based trust management for effective collaborative intrusion detection networks. *IEEE Transactions on Network and Service Management* 8 (2): 79–91.

29 Steven, C. (2011). Securing collaborative intrusion detection systems. *IEEE Security & Privacy* 9 (6): 36–42.

30 Gu, Q., Zang, W., Yu, M., and Liu, P. (2012). Collaborative traffic-aware intrusion monitoring in multi-channel mesh networks. *11th IEEE International Conference on Trust, Security and Privacy in Computing and Communications, TrustCom 2012, Liverpool,* United Kingdom (25–27 June 2012), pp. 793–800. IEEE Computer Society.

31 Pérez, M.G., Mármol, F.G., Pérez, G.M., and Gómez-Skarmeta, A.F. (2013). RepCIDN: A reputation-based collaborative intrusion detection network to lessen the impact of malicious alarms. *Journal of Network and Systems Management* 21 (1): 128–167.

32 Li, W., Meng, Y., and Kwok, L.-f. (2013). Enhancing trust evaluation using intrusion sensitivity in collaborative intrusion detection networks: feasibility and challenges. *9th International Conference on Computational Intelligence and Security, CIS 2013,* Emei Mountain, Sichan Province, China (14–15 December 2013), pp. 518–522. IEEE Computer Society.

33 Morais, A.M.P. and Cavalli, A.R. (2014). A distributed and collaborative intrusion detection architecture for wireless mesh networks. *MONET* 19 (1): 101–120.

34 Kumar, N. and Chilamkurti, N. (2014). Collaborative trust aware intelligent intrusion detection in VANETs. *Computers and Electrical Engineering* 40 (6): 1981–1996.

35 Cordero, C.G., Vasilomanolakis, E., Mühlhäuser, M., and Fischer, M. (2015). Community-based collaborative intrusion detection. In: *Security and Privacy in Communication Networks - 11th International Conference, SecureComm 2015, Dallas, TX, USA, October 26–29, 2015, Revised Selected Papers, Lecture Notes of the Institute for Computer Sciences, Social Informatics and Telecommunications Engineering,* vol. 164, 665–681. Springer.

36 Vasilomanolakis, E., Stahn, M., Cordero, C.G., and Mühlhäuser, M. (2015). Probe-response attacks on collaborative intrusion detection systems: effectiveness and countermeasures. *2015 IEEE Conference on Communications and Network Security, CNS 2015,* Florence, Italy (28–30 September 2015), pp. 699–700. IEEE.

37 Liu, X., Zhu, P., Zhang, Y., and Chen, K. (2015). A collaborative intrusion detection mechanism against false data injection attack in advanced metering infrastructure. *IEEE Transactions on Smart Grid* 6 (5): 2435–2443.

38 Andreolini, M., Colajanni, M., and Marchetti, M. (2015). A collaborative framework for intrusion detection in mobile networks. *Information Sciences* 321: 179–192.

39 Chen, X.-F. and Yu, S. (2016). A collaborative intrusion detection system against DDoS for SDN. *IEICE Transactions* 99-D (9): 2395–2399.

40 Fung, C.J. and Zhu, Q. (2016). FACID: A trust-based collaborative decision framework for intrusion detection networks. *Ad Hoc Networks* 53: 17–31.

41 Li, W., Meng, W., Kwok, L.-f., and Ip, H.H.-S. (2017). Enhancing collaborative intrusion detection networks against insider attacks using supervised intrusion sensitivity-based trust management model. *Journal of Network and Computer Applications* 77: 135–145.

42 Patel, A., Alhussian, H., Pedersen, J.M. et al. (2017). A nifty collaborative intrusion detection and prevention architecture for smart grid ecosystems. *Computers & Security* 64: 92–109.

43 Tug, S., Meng, W., and Wang, Y. (2018). CBSigIDS: Towards collaborative blockchained signature-based intrusion detection. *IEEE International Conference on Internet of Things (iThings) and IEEE Green Computing and Communications (GreenCom) and IEEE Cyber, Physical and Social Computing (CPSCom) and IEEE Smart Data (SmartData), iThings/GreenCom/CPSCom/SmartData 2018*, Halifax, NS, Canada (July 30 - August 3, 2018), pp. 1228–1235. IEEE.

44 Cordero, C.G., Traverso, G., Nojoumian, M. et al. (2018). *Sphinx*: a colluder-resistant trust mechanism for collaborative intrusion detection. *IEEE Access* 6: 72427–72438.

45 Hong, J. and Liu, C.-C. (2019). Intelligent electronic devices with collaborative intrusion detection systems. *IEEE Transactions on Smart Grid* 10 (1): 271–281.

46 Arshad, J., Azad, M.A., Abdellatif, M.M. et al. (2019). COLIDE: a collaborative intrusion detection framework for internet of things. *IET Networks* 8 (1): 3–14.

47 Meng, W., Li, W., and Zhu, L. (2020). Enhancing medical smartphone networks via blockchain-based trust management against insider attacks. *IEEE Transactions on Engineering Management* 67 (4): 1377–1386.

48 Hu, C., Liu, Z., Chen, Z., and Liu, B. (2006). Fuzzy optimization for security sensors deployment in collaborative intrusion detection system. In: *3rd International Conference on Fuzzy Systems and Knowledge Discovery, FSKD 2006, September 24–28, 2006, Proceedings*, (eds. Lipo Wang, Licheng Jiao, Guangming Shi, Xue Li, Jing Liu), *Lecture Notes in Computer Science*, vol. 4223, 743–752. Xi'an, China: Springer.

49 Fung, C.J., Baysal, O., Zhang, J. et al. (2008). Trust management for host-based collaborative intrusion detection. In: *Managing Large-Scale Service Deployment, 19th IFIP/IEEE International Workshop on Distributed Systems: Operations and Management, DSOM 2008, September 22-26, 2008. Proceedings*, (eds. Filip De Turck, Wolfgang Kellerer, George Kormentzas), *Lecture Notes in Computer Science*, vol. 5273, 109–122. Samos Island, Greece: Springer.

50 Marchetti, M., Messori, M., and Colajanni, M. (2009). Peer-to-peer architecture for collaborative intrusion and malware detection on a large scale. In: *Information Security, 12th International Conference, ISC, 2009, September 7-9, 2009. Proceedings*, (eds. Pierangela Samarati, Moti Yung, Fabio Martinelli,

Claudio Agostino Ardagna), *Lecture Notes in Computer Science*, vol. 5735, 475–490. Pisa, Italy: Springer.

51 Zhou, C.V., Leckie, C., and Karunasekera, S. (2009). Decentralized multi-dimensional alert correlation for collaborative intrusion detection. *Journal of Network and Computer Applications* 32 (5): 1106–1123.

52 Zhu, Q., Fung, C.J., Boutaba, R., and Basar, T. (2010). A distributed sequential algorithm for collaborative intrusion detection networks. *Proceedings of IEEE International Conference on Communications, ICC 2010*, Cape Town, South Africa (23–27 May 2010), pp. 1–6. IEEE.

53 Fung, C.J., Zhang, J., and Boutaba, R. (2010). Effective acquaintance management for collaborative intrusion detection networks. *Proceedings of the 6th International Conference on Network and Service Management, CNSM 2010*, Niagara Falls, Canada (25–29 October 2010), pp. 158–165. IEEE.

54 Grunewald, D., Chinnow, J., Bye, R. et al. (2011). Framework for evaluating collaborative intrusion detection systems. In: *Informatik 2011: Informatik schafft Communities, Beiträge der 41. Jahrestagung der Gesellschaft für Informatik e.V. (GI)*, 4.-7.10.2011, Berlin, Deutschland (Abstract Proceedings), *LNI*, (eds. Hans-Ulrich Heiß, Peter Pepper, Holger Schlingloff, Jörg Schneider) vol. P-192, 116. GI.

55 Li, W., Meng, W., and Kwok, L.-f. (2014). Design of intrusion sensitivity-based trust management model for collaborative intrusion detection networks. In: *Trust Management VIII - 8th IFIP WG 11.11 International Conference, IFIPTM 2014, Singapore (7–10 July 2014). Proceedings, IFIP Advances in Information and Communication Technology*, (eds. Jianying Zhou, Nurit Gal-Oz, Jie Zhang, Ehud Gudes) vol. 430, 61–76. Springer.

56 Li, W. and Meng, W. (2016). Enhancing collaborative intrusion detection networks using intrusion sensitivity in detecting pollution attacks. *Information & Computer Security* 24 (3): 265–276.

57 Chen, X.-F. and Yu, S.-Z. (2016). CIPA: A collaborative intrusion prevention architecture for programmable network and SDN. *Computers & Security* 58: 1–19.

58 Zhang, T. and Zhu, Q. (2018). Distributed privacy-preserving collaborative intrusion detection systems for VANETs. *IEEE Transactions on Signal and Information Processing Over Networks* 4 (1): 148–161. https://doi.org/10.1109/TSIPN.2018.2801622.

59 Zhou, Q. and Pezaros, D.P. (2019). BIDS: Bio-inspired, collaborative intrusion detection for software defined networks. *2019 IEEE International Conference on Communications, ICC 2019*, Shanghai, China (20–24 May 2019), pp. 1–6. IEEE.

60 Vasilomanolakis, E. and Mühlhäuser, M. (2019). Detection and mitigation of monitor identification attacks in collaborative intrusion detection systems. *International Journal of Network Management* 29 (2). 1–12.

61 Meng, W., Li, W., Tug, S., and Tan, J. (2020). Towards blockchain-enabled single character frequency-based exclusive signature matching in IoT-assisted smart cities. *Journal of Parallel and Distributed Computing* 144: 268–277.

62 Ma, Z., Liu, L., and Meng, W. (2020). Towards multiple-mix-attack detection via consensus-based trust management in IoT networks. *Computers & Security* 96: 101898.

63 Li, W., Meng, W., and Zhu, H. (2020). Towards collaborative intrusion detection enhancement against insider attacks with multi-level trust. In: *19th IEEE International Conference on Trust, Security and Privacy in Computing and Communications, TrustCom 2020*, Guangzhou, China (December 29, 2020 - January 1, 2021) (ed. G. Wang, R.K.L. Ko, Md.Z.A. Bhuiyan, and Y. Pan), 1179–1186. IEEE.

64 Li, W., Meng, W., Parra-Arnau, J., and Choo, K.-K.R. (2021). Enhancing challenge-based collaborative intrusion detection against insider attacks using spatial correlation. *IEEE Conference on Dependable and Secure Computing, DSC 2021*, Aizuwakamatsu, Japan (January 30 - February 2, 2021), pp. 1–8. IEEE.

65 Meng, W., Li, W., Yang, L.T., and Li, P. (2020). Enhancing challenge-based collaborative intrusion detection networks against insider attacks using blockchain. *International Journal of Information Security* 19 (3): 279–290.

66 Li, W., Meng, W., and Au, M.H. (2020). Enhancing collaborative intrusion detection via disagreement-based semi-supervised learning in iot environments. *Journal of Network and Computer Applications* 161: 102631.

67 Meng, W., Luo, X., Li, W., and Li, Y. (2016). Design and evaluation of advanced collusion attacks on collaborative intrusion detection networks in practice. *2016 IEEE Trustcom/BigDataSE/ISPA*, Tianjin, China (23–26 August 2016), pp. 1061–1068. IEEE.

68 Li, W., Meng, W., Kwok, L.-f., and Ip, H.H.-S. (2016). PMFA: Toward passive message fingerprint attacks on challenge-based collaborative intrusion detection networks. In: *Network and System Security - 10th International Conference, NSS 2016*, Taipei, Taiwan, September 28-30, 2016, Proceedings, *Lecture Notes in Computer Science*, (eds. Jiageng Chen, Vincenzo Piuri, Chunhua Su, Moti Yung) vol. 9955, pp. 433–449. Springer.

69 Li, W., Meng, W., Kwok, L.F., and Ip, H.H.S. (2018). Developing advanced fingerprint attacks on challenge-based collaborative intrusion detection networks. *Cluster Computing* 21 (1): 299–310.

70 Li, W., Meng, W., and Kwok, L.F. (2017). SOOA: Exploring special on-off attacks on challenge-based collaborative intrusion detection networks. In: *12th International Conference on Green, Pervasive, and Cloud Computing (GPC), Cetara, Italy, May 11-14, 2017*, Proceedings, *Lecture Notes in Computer Science*, (eds. Man Ho Allen Au, Arcangelo Castiglione, Kim-Kwang Raymond Choo, Francesco Palmieri, Kuan-Ching Li) vol. 10232, 402–415.

71 Li, W., Meng, W., and Kwok, L.F. (2018). Investigating the influence of special on-off attacks on challenge-based collaborative intrusion detection networks. *Future Internet* 10 (1): 6.

72 Meng, W., Li, W., Jiang, L. et al. (2019). Practical Bayesian poisoning attacks on challenge-based collaborative intrusion detection networks. In: *Computer Security - ESORICS 2019 - 24th European Symposium on Research in Computer Security, Luxembourg, September 23–27, 2019, Proceedings, Part I*, (eds. Kazue Sako, Steve A. Schneider, Peter Y. A. Ryan), *Lecture Notes in Computer Science*, vol. 11735, pp. 493–511. Springer.

73 Li, W., Meng, W., Wang, Y. et al. (2018). Identifying passive message finger-print attacks via honey challenge in collaborative intrusion detection networks. *17th IEEE International Conference On Trust, Security And Privacy In Computing And Communications / 12th IEEE International Conference On Big Data Science And Engineering, TrustCom/BigDataSE 2018*, New York, NY, USA (1–3 August 2018), pp. 1208–1213. IEEE.

74 Madsen, D., Li, W., Meng, W., and Wang, Y. (2018). Evaluating the impact of intrusion sensitivity on securing collaborative intrusion detection networks against SOOA. In: *18th International Conference on Algorithms and Architectures for Parallel Processing, Guangzhou, China, November 15–17, 2018, Proceedings, Part IV*, (eds. Jaideep Vaidya, Jin Li), *Lecture Notes in Computer Science*, vol. 11337, pp. 481–494. Springer.

75 Tuan, T.A. (2006). A game-theoretic analysis of trust management in P2P systems. *2006 1st International Conference on Communications and Electronics*.

76 Pérez, M.G., Mármol, F., Pérez, G.M., and Gómez-Skarmeta, A.F. (2014). Building a reputation-based bootstrapping mechanism for newcomers in collaborative alert systems. *Journal of Computer and System Sciences* 80 (3): 571–590.

77 Vasilomanolakis, E., Habib, S.M., Milaszewicz, P. et al. (2017). Towards trust-aware collaborative intrusion detection: challenges and solutions. In: *Trust Management XI - 11th IFIP WG 11.11 International Conference, IFIPTM 2017, June 12-16, 2017, Proceedings, IFIP Advances in Information and Communication Technology*, vol. 505, 94–109. Gothenburg: Springer.

78 Duma, C., Karresand, M., Shahmehri, N., and Caronni, G. (2006). A trust-aware, P2P-based overlay for intrusion detection. *17th International Workshop on Database and Expert Systems Applications (DEXA 2006)*, Krakow, Poland (4–8 September 2006), pp. 692–697. IEEE Computer Society.

79 Habib, S.M., Volk, F., Hauke, S., and Mühlhäuser, M. (2015). Computational trust methods for security quantification in the cloud ecosystem. In: *The*

Cloud Security Ecosystem - Technical, Legal, Business and Management Issues, 463–493. Elsevier.

80 Sahay, R., Meng, W., Estay, D.A.S. et al. (2019). CyberShip-IoT: A dynamic and adaptive SDN-based security policy enforcement framework for ships. *Future Generation Computer Systems* 100: 736–750.

81 Xu, Q., Su, Z., and Yang, Q. (2020). Blockchain-based trustworthy edge caching scheme for mobile cyber-physical system. *IEEE Internet of Things Journal* 7 (2): 1098–1110.

82 Li, W., Meng, W., Su, C., and Kwok, L.F. (2018). Towards false alarm reduction using fuzzy if-then rules for medical cyber physical systems. *IEEE Access* 6: 6530–6539.

83 Meng, W., Li, W., Yang, L.T., and Li, P. (2019). Enhancing challenge-based collaborative intrusion detection networks against insider attacks using blockchain. *International Journal of Information Security*. https://doi.org/10.1007/s10207-019-00462-x.

84 Milijic, M. (2019). 29+ smartphone usage statistics: around the world in 2020. https://leftronic.com/smartphone-usage-statistics/ (accessed 24 September 2021).

85 Meng, W., Fei, F., Li, W., and Au, M.H. (2017). Evaluating challenge-based trust mechanism in medical smartphone networks: an empirical study. *2017 IEEE Global Communications Conference, GLOBECOM 2017*, Singapore (4–8 December 2017), pp. 1–6. IEEE.

86 Bu, S., Yu, F.R., Liu, P.X. et al. (2011). Distributed combined authentication and intrusion detection with data fusion in high-security mobile ad hoc networks. *IEEE Transactions on Vehicular Technology* 60 (3): 1025–1036.

87 Chekired, D.A., Khoukhi, L., and Mouftah, H.T. (2019). Fog-based distributed intrusion detection system against false metering attacks in smart grid. *2019 IEEE International Conference on Communications, ICC 2019*, Shanghai, China (20–24 May 2019), pp. 1–6. IEEE.

88 Shekari, T., Bayens, C., Cohen, M. et al. (2019). RFDIDS: radio frequency-based distributed intrusion detection system for the power grid. *26th Annual Network and Distributed System Security Symposium, NDSS 2019*, San Diego, California, USA (24–27 February 2019). The Internet Society.

89 Ali, R., Kim, B.-S., Kim, S.W. et al. (2020). (ReLBT): A reinforcement learning-enabled listen before talk mechanism for LTE-LAA and Wi-Fi coexistence in IoT. *Computer Communications* 150: 498–505.

90 Wang, Y., Meng, W., Li, W. et al. (2019). Adaptive machine learning-based alarm reduction via edge computing for distributed intrusion detection systems. *Concurrency Computation Practice and Experience* 31 (19). 1–12.

6

Cyber-Securing IoT Infrastructure by Modeling Network Traffic

Hassan Habibi Gharakheili, Ayyoob Hamza, and Vijay Sivaraman

School of Electrical Engineering and Telecommunications, University of New South Wales, Sydney, Australia

6.1 Introduction

New Internet-of-Things (IoT) devices are emerging each week in the market, and many of them are ridden with security holes, as amply demonstrated by many research groups including ours [1, 2]. The growing use of IoT devices in smart environments (homes, buildings, enterprises, and cities) not only poses grave risks to the security and privacy of personal and organizational data gleaned from sensors [3] but also provides a launching pad for attacks to systems within [4] and outside [1] the organization. Current approaches to IoT security evaluation are piecemeal and ad hoc.

Though many research papers (including ours [1, 3]) have identified myriad security flaws in IoT devices, few have suggested solutions beyond the patching of these flaws by the respective manufacturers, which is doomed to failure, given the large number of vendors and their limited motivation to care for a device beyond its sale. A promising direction is to monitor and lockdown the network activity of IoT devices to detect and block misbehavior [5, 6], giving the network operator a second line of defense against compromised devices without relying solely on appropriate security protections from each vendor. Success of this approach relies on knowing the expected network behavior of each IoT device, and the interactions of these devices in a specific deployment environment such as a building. Luckily, frameworks for formal specification of these two aspects are starting to emerge – the IETF (internet engineering task force) MUD (Manufacturer Usage Description) standard [7] provides a language for describing the network activity of an IoT device, while the Brick standard [8] provides a metadata schema for buildings including the interactions of the various components within.

Security and Privacy in the Internet of Things: Architectures, Techniques, and Applications,
First Edition. Edited by Ali Ismail Awad and Jemal Abawajy.

Network intrusion detection systems have been studied extensively by the research community and look either for *signatures* of known attacks or *anomalies* indicative of deviation from normal behavior. Nearly all deployed solutions, including software tools like Bro [9] and Snort [10], and commercial hardware appliances, belong to the former category. However, attack-signature-based solutions are difficult to scale cost-effectively: software solutions can barely do a few hundred Mbps per central processing unit (CPU) core, while specialized hardware appliances run into hundreds of thousands of dollars for a few tens of Gbps throughput. Further, this approach is increasingly under threat, as the percentage of network traffic that is encrypted increases and also has poor resilience to morphed attacks that can render known signatures useless.

Anomaly detection holds promise as a way of detecting new and unknown threats, but despite extensive academic research (see [11] for a survey) has had very limited success in operational environments. The reasons for this are manifold [12]: "normal" network traffic can exhibit much more diversity than expected; obtaining "ground truth" on attacks in order to train the classifiers is difficult; evaluating outputs can be difficult due to the lack of appropriate datasets; false positives incur a high cost on network administrators to investigate; and there is often a semantic gap between detection of an anomaly and actionable reports for the network operator.

Arguably, the most significant challenge to the design and evaluation of data analytics techniques for cybersecurity is the lack of access to large volumes of real and contemporary data. Too many research studies have relied on the open defense advanced research projects agency (DARPA) [13] and knowledge discovery in databases (KDD) [14] datasets, which are synthetic, outdated, and even questionable in their attack patterns [15]. By contrast, in this chapter we will rely on real data obtained from our university campus network [16] and IoT testbed of about 30 different types of real IoT devices, constituting a dataset that is current, large, and diverse.

In recent years, there has been growing recognition that machine learning is not a "silver bullet" that can magically detect all cyber-security threats, but is instead more effective in a narrower context – when the threat model is clear and the scope of the target activity is narrow. There have been recent successes in applying machine learning to Botnet blocking [17], distributed denial of service (DDoS) detection [18], and encrypted malware classification [19]. We will demonstrate later in this chapter, how our approach extracts flow-level attributes from the IoT network using software-defined networking techniques, enabling us to develop a more sophisticated breed of data-driven models that can identify (and eventually mitigate) a wide range of cyber-attacks.

Our specific contributions are as follows. We begin by highlighting IoT network threats and attack vectors, as well as existing countermeasures and their limitations (Section 6.2). Next, we develop a systematic approach to model the network behavior of IoT devices (Section 6.3), automatically enforcing their behavior and monitoring real-time activity using a set of flow-based anomaly detectors.

6.2 Cyber-Attacks on IoT Infrastructure

There are several public reports [20] about many sophisticated cyber-attacks on poorly secured IoT devices. Many IoT networks have proven vulnerable to both active and passive attacks by leaking private information [21], allowing unauthorized access [22], and being open to denial of service attacks [1].

In this chapter, we begin by highlighting various types of network attacks on IoT infrastructure, and next discuss their techniques and impact. IoT attacks can be viewed in various dimensions such as their target pillar (i.e. confidentiality, integrity, and availability) of information security, or their target layer of the Internet protocol stack (i.e. application, transport, and data-link). We, in this chapter, are particularly interested in the impacts of IoT cyber-threats. Network-based attacks on IoT infrastructure can be launched either in passive or active mode, as shown in Figure 6.1. Passive attackers often do not leave any trace of their activity on the network and aim to collect information from the environment by listening to network communications. On the other hand, active attackers

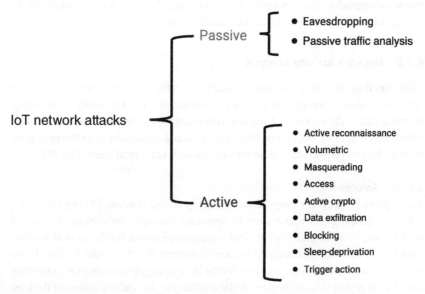

Figure 6.1 Taxonomy of IoT network attacks.

generate network traffic (directly or indirectly) to their target devices. In what follows, we discuss each of these network attacks individually.

6.2.1 Eavesdropping

Certain attackers tend to eavesdrop by secretly listening to potential information exposed on the network that enables them to profile a subject device or the entire network. Consumer IoT devices often expose their identity by way of broadcasting on the network via protocols such as simple service discovery protocol (SSDP) and multicast domain name service (mDNS), indicating their type, firmware, or even certain configurations. Sivaraman et al. [4] demonstrated how malicious entities can use the collected information to remotely control the vulnerable device. Also, many IoT devices communicate unencrypted traffic with mobile devices on the local network [23], relying upon over-the-air encryption provided by wireless mediums. However, the vulnerability in Wi-Fi chips has led attackers breaking encryption protocols such as WPA2-Personal and WPA2-Enterprise and accessing the device information [24] use this information to compromise, degrade, or disrupt networks.

6.2.1.1 Solutions

Eavesdropping leaves no trace and hence is challenging to detect. The only defense is to prevent these attacks by use of standard cryptography and enabling regular firmware upgrades. Manufacturers need to ensure that they use "recommended" cipher suites for all local and Internet communications.

6.2.2 Network Activity Analysis

Attackers may choose to passively capture the traffic of a device and/or group of network devices. This can be done either within the local network or remotely from outside of the network. A rogue access point is an example of these attacks by which users' activity, the composition of connected devices, or even the operating state of devices (on/off) can be deduced even with encrypted traffic [25, 26].

6.2.2.1 Solutions

These privacy risks can only be prevented by traffic shaping [27] or tunneling [25]. The shaping technique aims to obfuscate network traffic flows by way of payload padding, fragmentation, and randomized cover traffic so that various user/device activities get mapped to a traffic pattern that is not identifiable. Datta et al. [27] developed a python library for traffic shaping at a reasonable computing cost, but applying this method might be challenging for battery-powered devices. Tunneling technique, Apthorpe et al. [25], on the other hand, aims to aggregate all traffic flows into a single bidirectional flow, reducing the chance of inferencing for intruders.

6.2.3 Active Reconnaissance

Reconnaissance attacks launched on the network are harmless to the IoT device and its functionality, but they can lead to more significant attacks on the subject device and/or other network entities. Scanning IP addresses or transport control protocol (TCP)/user datagram protocol (UDP) ports, or sending SSDP search queries are few techniques that have been used on IoT devices [4, 28–30]. There exist certain online engines like Shodan [29] that actively probe the entire space of public IP addresses, looking for vulnerable open ports. Botnets (e.g. Mirai botnet) also scan the Internet to identify open telnet ports, leading to large-scale distributed attacks [28]. SSDP search queries have been used for device fingerprinting, which enables illegitimate remote access [4, 30].

6.2.3.1 Solutions

To defend against active reconnaissance attacks, one can use techniques such as rate limiting, activity modeling, or access controls. Mehdi et al. [31] proposed to track new connections and rate limit those sources which perform port scans, and Odhaviya et al. [32] used TCP reset (RST) packets as an indicator of port scans. IP scans can only be detected by thresholding [33] which can be defeated by slow-rate stealth scans. graph-based intrusion detection system (GrIDS) [34] proposed graph-based activity modeling which tracks connected hosts as nodes and their inter-communications as edges of the graph. The graph data structure provides a holistic view of the entire network and its dynamics, which allows for performing a variety of algorithms to detect attacks. Porras and Valdes [35] proposed a statistical model for individual connected devices that captures attributes of their short-term and long-term communications. For example an attribute is the count of SYN packets – a sudden rise in the count indicates a deviation from the normal state. Activity modeling can detect scans but will incur a significant cost of computing to maintain states of network devices and their communications. Lastly, access controls (whitelisting the authorized network endpoints and/or intended service protocols) can effectively limit the attack surface. For example, Demetriou et al. [36] proposed an Software-Defined Networking (SDN)-based architecture that enables network administrators to authorize their devices before they can interact with each other. Also, specifications of MUD [7] can restrict network communications of each connected IoT device to a limited set of services that are allowed (and declared by the manufacturer) to use. Any communication with disallowed (or unspecified) services is considered a threat [37]. However, certain attacks which conform to the allowed services cannot be detected.

6.2.4 Volumetric Attack

Recent reports [38] show that attackers continue to exploit insecure IoT devices to launch volumetric attacks in the form of DoS, DDoS, brute force, and TCP SYN/UDP flooding. These attacks can target either a remote Internet-based server or the IoT device itself to make them inaccessible [39]. Moreover, the progression of botnets [40, 41] such as Mirai and Persirai, infecting millions of IoT devices, is enabling destructive cyber-campaigns of unprecedented magnitude to be launched. Mirai botnet was the first and most notable IoT-based attack which brought down many Internet services in 2016, highlighting the impact of IoT vulnerability. Since then some other botnets such as Persirai [42], Bashlite [43], Okiru [44], wicked [45], and Torii [46] have emerged. Botnet-based attacks typically start by hijacking many insecure IoT devices and coordinate with a command-and-control server to launch a volumetric DDoS have emerged on their victim. In certain settings, DDoS attacks may be generated by getting reflected and amplified traffic from IoT devices [1].

6.2.4.1 Solutions

Detecting volumetric attacks on computer networks and systems has been extensively studied by the research community. Existing methods primarily look for either *signatures* of known attacks or *anomalies* indicative of deviation from normal behavior. Signature-based detection is commonly adopted by enterprise security appliances and requires prior knowledge of IoT threats and their signature (mostly at packet level). This approach can incur significant computing cost due to inspection of individual packets and also may miss [37] zero-day (new) attacks. We showed in our earlier work [39] that signature-based tools are only able to detect a limited number of volumetric attacks (on IoT devices).

On the other hand, there are many studies on employing either entropy-based [31, 47–49] or machine learning-based [50–52] techniques to detect new volumetric attacks. The entropy-based approach is primarily used for detecting types of volumetric attacks that generate a large number of flows. Authors of [47–49] use sample entropy of source/destination IP addresses and port numbers to determine whether a large variation (above a certain threshold) from the norm is observed in the network traffic. In [47, 48] and [49], the authors have applied this technique to detect attacks in ISP networks, backbone networks, and campus networks, respectively. Once an anomaly is detected, identifying the exact fine-grained attack flows (three-tuple, four-tuple, or five-tuple) is challenging at scale since this method computes a single entropy value for the measured traffic data, and hence pinpointing the cause of attack becomes nontrivial without maintaining all possible states (infeasible in a large network).

Works in [50–52] develop two-class classifiers to distinguish benign from attack traffic. Authors of [50, 51] use traffic features like flow-level statistics (i.e. packet/byte count and duration), percentage of bidirectional flows, the growth rate of unidirectional flows, and the growth rate of unique ports for training their classifiers. Work in [52] employed deep learning algorithms with a similar set of features to classify normal and abnormal traffic. This model was applied to IoT devices by authors of [53]. However, their evaluation is limited to simulated traffic in the Mininet emulator environment that does not represent the behavior of real IoT devices.

6.2.5 Masquerading Attack

In this type of attack, the source or content of network traffic is altered or spoofed without being easily discovered by the victim devices. Consumer IoT devices often use weak ciphers [54], weak encryption, or no encryption in their network traffic. Using a simple DNS spoofing attack, the attacker can make the device trust a fake server, leading to data leaks and compromising the privacy of users/businesses. Moreover, protocols such as Network Time Protocol (NTP) are widely used in IoT devices. Spoofing the time in NTP packets can make devices fail to verify their (expired) SSL/TLS (transport layer security) certificates, which can render them inactive. Routing information attacks can be launched on devices to take them offline with very few spoofed packets. Address resolution protocol (ARP) spoofing can result in man-in-the-middle attacks. Note that IoT devices have no embedded agent to provide their operational states, and thus it becomes challenging for network administrators to deduce or identify the root cause of such failures.

6.2.5.1 Solutions

There is no generic solution for detecting all types of masquerading attacks – a custom strategy is required for each attack type. However, these attacks can be detected by profiling normal activities or prevented by the use of secured communication protocols.

Attacks like ARP or IP spoofing can be detected by inspecting packets and verifying the network profile [55, 56] – maintaining an ARP table or map of IP to media access control (MAC) addresses. For NTP spoofing attacks, time offsets can be inspected to detect large deviations [57]. Detecting DNS spoof requires knowledge of all blacklisted or whitelisted IP addresses [58]. Activity profiling can incur heavy computing costs in high throughput networks due to deep packet inspection.

Using secured network protocols can limit the attack surface to some extent. For example, domain name system security extensions (DNSSEC) can protect IoT devices from DNS spoofing [59]. It was shown in [23] that none of the consumer IoT devices examined had DNSSEC enabled. Also, NTP spoofing attacks can be limited by the use of NTP v4 which limits the variation of NTP offset value, but slow-changing NTP offset can still be harmful. Lastly, to reduce the impact of such attacks, it is recommended to use strong cipher suites for all communication flows.

6.2.6 Access Attack

Many consumer IoT devices have been reported to have weak authentication in their local services, which has led to many replay attacks and unauthorized accesses [23, 60]. This issue has been more pronounced due to the use of default passwords in commercial IoT devices [61, 62].

The largest botnet attack, Mirai, was instigated because of having open Telnet ports in IoT devices. Devices such as LiFX bulb and TPLink bulb with no authentication in local services, allowed attackers to take control of them by either replaying or crafting prerecorded control messages [3]. The payload structure of control messages can be determined by passively recording network traffic or from the device open application programming interfaces (APIs). We found in our lab that Genbolt IP cameras expose an insecure web portal through which attackers can access their firmware and enable Telnet service with root privileges. In another example, arbitrary YouTube videos can be fed on and played on Chromecast with no authentication [3]. Manufacturers often allow unprotected (insecure) local services on consumer devices for ease of use, expecting network routers to have mechanisms for restricting network-based access. However, work in [4] demonstrated that devices can be accessed by way of port forwarding on home routers, which is enabled by malware embedded into a mobile application. Also, authors in [30] showed that local services can be accessed through web browsers using DNS rebinding attacks.

6.2.6.1 Solutions

These attacks can be limited by applying access controls or resistant schemes. Traditional network access controls can be used to allow only authorized devices. Work in [36] proposed a mobile agent for home users to specify authorized applications originating from individual IoT devices on the network. This solution is difficult to apply in practice because of the need for users to install a mobile application. Therefore, more sophisticated models are required to protect devices from such attacks originating from authorized devices. One possible way is to model the user interactions with their authorized devices; yet capturing the ground-truth data is very challenging. Another approach is the use of password protection for all services. Applying rate-limiting for password retries or using session-id for individual message transactions [63] would limit the attack surface.

6.2.7 Active Crypto Attack

The use of strong ciphers is the first step to achieving security, but many IoT devices are found [54] to not implement best-practice guidelines, and thus integrity and privacy will be at risk. Ciphers are considered to be one of the fundamental requirements for information security. Work in [64] demonstrated attacks on intimacy devices by way of man-in-the-middle hijacking the user session. This type of attack uses SSL split tools to break an existing TLS connection and takes advantage of the device vulnerability in not verifying the trustworthiness of provided certificate. In another example, we have found some of the consumer IP cameras (manufactured by D-Link, Joodan, and Genbolt) do not encrypt their transmitted data, enabling man-in-the-middle intruders to capture images that the camera transmits its the mobile app.

6.2.7.1 Solutions

The only prevention is to follow best-practice security guidelines in using strong and recommended cipher suites, for example use of TLSv2 algorithm; validating CA certificates with a trusted store; or checking the expiry date of certificates [65]. A network security solution may be able to determine whether the cipher suites used by connected IoT devices are strong and recommended. However, verification of certificate by the device cannot be passively determined – such test requires an active assessment of the device by exposing it to fake certificates.

6.2.8 Data Exfiltration

Exfiltration of sensitive data is of concern for a variety of organizations such as financial organizations, universities, or insurance companies. Attackers manage to access sensitive data records on organizational servers and transfer them via covert channels to their command-and-control servers. There are many methods to exfiltrate data, and HTTP, FTP, DNS, secure shell (SSH), and Email are a few examples of those [66, 67].

Researchers in [68] demonstrated an attack where they collect sensitive documents that are printed using 3D printers. Exfiltrated data can lead to ransomware attacks [69]. Interestingly, work in [70] demonstrated a novel type of exfiltration attack by transmitting the collected data outside by way of changing the intensity of smart lightbulb in the home network – an outside receiver decodes the data.

6.2.8.1 Solutions

There exist a number of works on detecting exfiltration attacks on general-purpose devices by applying machine learning techniques. For example, authors of [71] were able to detect DNS-based exfiltration attack by modeling the pattern of benign domain names; anything outside of the normal pattern is marked as malicious. That said, sophisticated exfiltration attacks can be launched over encrypted channels like HTTPS or secure file transfer protocol (SFTP), which require more attention from the research community.

6.2.9 Blocking Attack

Physical damage, jamming, or destructing the functionality of the device falls under the blocking attack category. They can lead to severe consequences. Examples of these are an attacker who targets to disable temperature sensors in a data center, or disables motion sensors in critical infrastructure. Many IoT specialized platforms monitor the health of their associated devices in production, but these real-time insights are often not made available to the network administrators.

The intention of blocking attacks is to prevent the device to communicate with its intended controllers. IoT devices are often installed and left unattended, and hence blocking remains unnoticeable unless a health monitoring service detects it. IoT platforms monitor the health of a device from its keep-alive notifications or published data. Purely monitoring the keep-alive notifications can be vulnerable since an adversarial attacker can selectively allow only health-check service communications, but block others.

6.2.9.1 Solutions

Different blocking attacks can be identified by physical inspection, monitoring the wireless channel, or monitoring the network activity. Physical inspection may not be fully accurate and does not scale cost-effectively as it is difficult to be automated and demands a human workforce. Monitoring wireless channels can be used to identify potential jamming frequencies. Jamming attacks are launched by generating noise to impact the signal-to-noise ratio of the receivers [72]. To avoid such wireless-based attacks, two approaches can be used: (i) enabling IoT hardware to differentiate noises from signals, but it can be energy-intensive and challenging to detect when the noise frequency changes rapidly [73], and (ii) using external noise detectors to identify the increase in noise levels, which can be fairly expensive for larger IoT deployments. Lastly, monitoring network activities can be considered as an automatic way to detect blocking attacks. A blocking attack would eventually impact the network behavior of subject devices. In our recent work [74], we automatically grouped IoT devices based on their physical location and modeled their benign network behavior using one-class classifier models.

6.2.10 Sleep Deprivation Attack

Battery-powered IoT devices are programmed by manufacturers to automatically switch to sleep mode (when inactive) in order to save energy and extend their longevity. However, attackers may target these devices by preventing them from sleeping. The attacker aims to keep the device active by consistently sending network packets to process, which eventually leads to battery exhaustion and making the device offline.

6.2.10.1 Solutions

The frequency (rate) of attack packets can help the detection solution. Since high-rate attacks (volumetric) can be detected relatively easier, sophisticated attackers may go to stealth mode, generating slow-paced persistent malicious traffic to their victim [75]. Modeling energy levels and resource usage of a device is an approach to identify such low-profile attacks [76]. At the network level, one may employ a number of approaches such as access control, rate limiting, or traffic modeling. Access controls can be used to limit the attack surface by allowing the endpoints which are expected to communicate with individual IoT devices on the network; however, attacks from authorized devices will not be detected. Authors of [77] proposed rate-limiting to prevent such attacks. However, identifying a single threshold (using historical network data) for various device types can be practically challenging. Lastly, modeling the sleep pattern would be an effective approach by which anomalies can be detected during the attack. Again, this modeling requires a decent amount of data on device traffic with ground-truth labels of their states (sleep, active, and transitions between sleep and active).

6.2.11 Trigger Action Attack

Consumer IoT devices can be integrated with each other by trigger-and-action rules. Considering a fire alarm with two operational states ("on" or "off") and a temperature sensor with three states ("hot," "cold," or "normal"), consumers may create a variety of trigger-and-action rules using the states of these devices to better manage their smart environment in an automated fashion. For example, if the temperature is high, then turn on the air conditioner (AC) and close the window. However, attackers can exploit such complex event transitions. For example they may attempt to increase the electricity consumption by turning on the AC, while spoofing the state of windows and temperature sensors. Authors of [78, 79] showed how exploiting trigger-and-action rules can lead to DoS and privacy attacks.

6.2.11.1 Solutions

There exist preliminary studies on developing solutions to address this specific attack vector using network data. Authors of [80] developed a machine learning-based method to identify the operating state of devices using their network activity behavior. This work considers three generic states, namely "idle," "active," and "boot" for connected devices. In a different approach, Apthorpe et al. [25] demonstrated how motion can inferred from encrypted network traffic [25]. However, identifying the state of a multitude of devices at scale can be practical. It is primarily due to a lack of annotated datasets and obfuscated communications.

6.3   Network Behavioral Model of IoTs

The current practice for securing organizational networks is to rely on traditional signature-based Intrusion Detection Systems (IDS) that inspect network traffic to detect attacks. However, such solutions are either extremely expensive if they are hardware-based, or unscalable to high data-rates if they are software-based. Further, the myriad variety of IoT devices, each with its own specific behavior and security vulnerabilities, makes it challenging for the IDS to distinguish normal from abnormal traffic that could be symptomatic of an attack.

IETF recently standardized [7] the MUD grammar and mechanism for specifying IoT device behavior that is gaining increasing interest from industry. IoT devices generally perform a specific function and therefore have a recognizable communication pattern [81], which can be captured formally and succinctly as a MUD profile.

This framework requires manufacturers of IoTs to publish a behavioral profile of their device, as they have the best knowledge of how their device is expected to behave when installed in a network; for example an IP camera may need to use DNS and DHCP on the local network and communicate with NTP servers and a specific cloud-based controller in the Internet, but nothing else. Such device behavior is manufacturer-specific. Knowing each device's intended behavior allows network operators to impose a tight set of access control list (ACL) restrictions per IoT device in operation, reducing the potential attack surface on their network.

Using the SDN paradigm, this formal behavioral profile can be translated to static and dynamic flow rules that can be enforced at run-time by the network – traffic that conforms to these rules can be allowed, while unexpected traffic inspected for potential intrusions. Such an approach dramatically reduces the load on the IDS, allowing it to scale in performance and identify device-specific threats.

The ability to control inappropriate communication between devices in the form of ACLs is expected to limit the attack surface on IoT devices; however, little is known about how MUD policies will get enforced in operational networks, and how they will interact with current and future intrusion detection systems.

6.3.1   Enforcing MUD Profile to Network

In this section, we outline our SDN-based system to enforce MUD policies and dynamically inspect exception traffic (nonconforming to MUD profile) which is a small fraction of total packets to/from IoT devices. The IETF MUD is a new standard. Hence, IoT device manufacturers have not yet provided MUD profiles for their devices. Therefore, in our previous work [82], we developed a tool - *MUDgee* - which automatically generates a MUD profile for an IoT device from its traffic trace

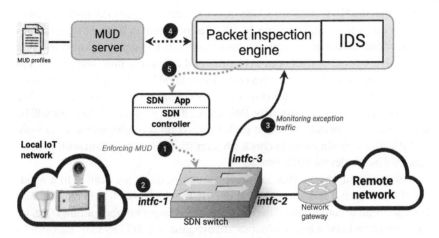

Figure 6.2 System architecture of enforcing and monitoring MUD behavior.

in order to make this process faster, easier, and more accurate. Our SDN-based system uses as input MUD profiles [83] of consumer IoT devices that we automatically generated by the MUDgee tool.

Figure 6.2 shows the functional blocks in the system architecture enforcing and monitoring MUD behavior, applied to a typical home or an enterprise network. IoT devices on the left can communicate with the local network as well as with remote servers (on-premise and/or Internet cloud) via a gateway. This system comprises a switch whose flow-table rules are managed dynamically by the SDN controller, a packet inspection engine, and a signature-based IDS.

The switch is initially configured by a default rule to mirror all traffic to the inspection engine on intfc-3, as shown by step ①. Packets from an IoT device, that has not yet been discovered, are forwarded (on intfc-2) and mirrored (on intfc-3), as shown by step ②. The inspection engine keeps track of already discovered devices on the local network (by maintaining a table of known MAC or IP addresses). Upon connection of a new device, its traffic is mirrored which helps the inspection engine detect this device and obtain the MUD URL from its initial packets (e.g. DHCP request) shown by step ③ in Figure 6.2. Thereafter, the inspection engine fetches the corresponding MUD profile from a MUD file server as shown by step ④. The MUD profile is stored (till its validity period), and its ACEs will be translated into a set of flow rules ("proactive"). Proactive rules are then inserted into the switch via the SDN controller as shown by step ⑤. ACEs can be directly translated to flow rules, but they require priority which is not captured by the current MUD standard. It is important to note that the order of flows becomes important when generating flow rules from a device MUD profile for preventing unwanted traffic to/from the device.

Note that the MUD standard allows manufacturers to specify Internet endpoints by their domain-name. Therefore, MUD ACEs pertinent to Internet/remote communications (with domain-name) cannot be directly translated to flow rules. This means that DNS responses need to be inspected in order to find their bindings at run-time and store them in a DNS cache (maintained by the inspection engine). Also, all Internet traffic of individual IoT devices are mirrored to check whether their remote IP address exists in the DNS cache; if yes, a "reactive" flow rule will be inserted into the switch. Note that packet inspection is conducted for flows associated with a domain name to check whether the flow is compliant with policies specified by the device MUD profile.

It is important to note that IoT local communications are limited to a handful of foundational flows (e.g. ARP and DHCP) which are often static, whereas Internet/remote communications are carried by a larger number of flows (e.g. up to 50 for some IoTs) whose endpoints are dynamic (e.g. HTTPS to h.canaryis.com). Therefore, flow rules corresponding to local communications remain permanent on the SDN switch, and an idle-timeout is set for each of the reactive flow rules of Internet communications for efficient use of limited TCAM inside the SDN switch.

As an example, Figure 6.3 shows flow rules translated from the MUD profile of the Canary camera. Shaded rows correspond to a snapshot of reactive flow rules since they vary over time. Domain-names are shown for Internet-based source/destination to make it easier to visualize (in actual flow-table, IP addresses are used). It is seen that the camera communicates with five subdomains of canaryis.com – one is HTTP, and five are HTTPS. Nonhighlighted rows correspond to proactive rules and the default rule (at the very bottom of Figure 6.3). As discussed earlier, proactive rules IV, XX, and XXI, respectively, mirror DNS replies as well as outgoing/incoming Internet traffic for each IoT device. Note that reactive rules would have a priority slightly higher than that of flows mirroring Internet traffic, but lower than of the DNS reply flow. This way, we stop mirroring packets of Internet flows that are correctly captured by the inspection engine.

Lastly, we see essential rules for basic operations over the network such as rules I, II, and III, respectively correspond to EAPoL (Extensible Authentication Protocol over LAN) packets, DHCP requests, and internet control message protocol (ICMP) replies from default gateway; rules V and VI relate to direct communication between the device and the default gateway for NTP communications; rules XXII and XXIII specifically match on ARP packets from/to the device.

6.3.2 MUD Protection Against Attacks

We analyze the efficacy of MUD profiles considering four categories of attacks: To/From Internet, and To/From local network. Details of our analysis are shown

flow-id	sEth	dEth	typeEth	Source	Destination	proto	sPort	dPort	priority	action
I	<devMAC>	*	0x888e	*	*	*	*	*	20	forward
II	<devMAC>	FF:FF:FF:FF:FF:FF	0x0800	*	*	*	*	67	20	forward
III	<gwMAC>	<devMAC>	0x0800	*	*	1	*	*	20	forward
IV	<devMAC>	<devMAC>	0x0800	*	gateway IP	17	*	53	20	forward
V	<gwMAC>	<devMAC>	0x0800	gateway IP	*	17	123	*	20	forward
VI	<devMAC>	<gwMAC>	0x0800	*	gateway IP	17	*	123	20	forward
VII	<gwMAC>	<devMAC>	*	*	*	17	53	*	20	forward & mirror
VIII	<gwMAC>	<devMAC>	0x0800	h.canaryis.com	*	6	80	*	11	forward
IX	<devMAC>	<gwMAC>	0x0800	*	h.canaryis.com	6	*	80	11	forward
X	<gwMAC>	<devMAC>	0x0800	h.canaryis.com	*	6	443	*	11	forward
XI	<devMAC>	<gwMAC>	0x0800	*	h.canaryis.com	6	*	443	11	forward
XII	<gwMAC>	<devMAC>	0x0800	o.canaryis.com	*	6	443	*	11	forward
XIII	<devMAC>	<gwMAC>	0x0800	*	o.canaryis.com	6	*	443	11	forward
XIV	<gwMAC>	<devMAC>	0x0800	b.canaryis.com	*	6	443	*	11	forward
XV	<devMAC>	<gwMAC>	0x0800	*	b.canaryis.com	6	*	443	11	forward
XVI	<gwMAC>	<devMAC>	0x0800	i.canaryis.com	*	6	443	*	11	forward
XVII	<devMAC>	<gwMAC>	0x0800	*	i.canaryis.com	6	*	443	11	forward
XVIII	<gwMAC>	<devMAC>	0x0800	m.canaryis.com	*	6	443	*	11	forward
XIX	<devMAC>	<gwMAC>	0x0800	*	m.canaryis.com	6	*	443	11	forward
XX	<devMAC>	<gwMAC>	0x0800	*	*	*	*	*	10	forward & mirror
XXI	<gwMAC>	<devMAC>	0x0800	*	*	*	*	*	10	forward & mirror
XXII	*	<devMAC>	0x0806	*	*	*	*	*	7	forward
XXIII	<devMAC>	*	0x0806	*	*	*	*	*	7	forward
XXIV	*	*	*	*	*	*	*	*	1	forward & mirror

Figure 6.3 Flow rules of Canary camera.

in Figure 6.4. For ease of visualization, we use color codes to indicate the device protection: dark grey for being secure, bright for moderately secure, and light grey for being insecure.

6.3.2.1 To Internet

Many IoT devices have been compromised due to insufficient (or even a lack of) authentication enforcement and have been employed in large-scale botnet attacks to popular Internet servers [38]. Mirai, Brickerbot, and Hajime are examples of major IoT-botnet-based attacks that were launched [40]. MUD helps isolate exception packets related to any attempts for botnet injection only if the device access controls are tightly defined, or the device does not expose Telnet or SSH services. In case of already compromised devices, MUD is again able to isolate exception packets related to attack flows that are not specified by the device profile.

We observe that the MUD policy for five devices (i.e. August camera, Belkin netcam, Ring doorbell, Samsung camera, and TPLink camera) cannot be tightly defined [82]. These devices allow peer-to-peer communications for streaming video to their corresponding mobile App (by way of STUN or a streaming server to initiate the handshake). These types of Internet communications necessitate the device to access an arbitrary range of IP addresses and port numbers, thus making the device vulnerable. Therefore, these devices are marked by light grey cells (in Figure 6.4 under *To-Internet* heading) for their insecurity against either TCP- or UDP-based attacks.

IoT devices have been used as reflectors to amplify TCP, UDP, or ICMP floods to Internet-based victims [1, 41]. Another observation is that all of our IoT devices are vulnerable to DNS spoofing, since they do not implement DNSSEC, thus not able to verify the integrity of reply packets. However, enforcing devices to use local DNS servers (e.g. in enterprise networks) can reduce the possibility of a DNS spoofing attack. Bright cells in Figure 6.4 indicate the DNS-related vulnerability. Overall, more than 80% of experimented IoT devices are moderately secure against reflection attacks on Internet-based victims.

6.3.2.2 From Internet

Attacks from the Internet are typically successful if perimeter security by NAT/firewall is not present, or malware exposes the local network to the Internet via port forwarding [4]. Shodan [29] runs frequent scanning to identify devices publicly accessible and publishes their IP address and open ports. Furthermore, Insecam [84] lists publicly available video camera feeds from various countries. If appropriate policies have been applied, then these unintended video feeds would have been blocked. Attackers, therefore, would be able to lookup these public repositories to hunt vulnerable IoT devices.

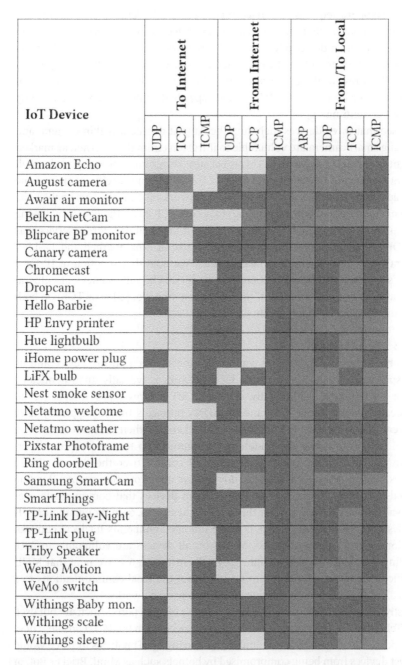

Figure 6.4 Protection of IoT devices (with MUD enforcement) against cyber-attacks.

MUD standard requires manufacturers to separately define local and Internet communications for their devices. This helps protect devices against port forwarding exposure – if the device offers a service to the local network, unintended remote access from the Internet can be prevented to some extent (excluding the spoofing of the IP address of the device's server). Moreover, unintended endpoints from Internet can be limited if the MUD policies specify the intended endpoints tightly (i.e. by a domain-name).

We see that three devices including the August doorbell, Belkin camera, and Samsung smart camera are vulnerable to intrusions from the Internet, as marked by light grey cells in Figure 6.4. This is because, for example the August doorbell camera is allowing Internet traffic from TCP 443 (with no specific domain-name or IP address). This device serves the local network on TCP 80. This enables the attacker (with help of port forwarding) to access TCP 80 on the device by having the source TCP port equals 443. We note that if the local or Internet service was on UDP (i.e. mismatched transport protocols like in Ring doorbell and TP-Link camera), this type of attack cannot succeed. We note that spoofing of IP for a device's server is still possible; therefore, all devices become somewhat vulnerable to attacks from the Internet if they talk same protocol (UDP or TCP) on both local and Internet channels, as shown by bright cells.

6.3.2.3 From/To Local

It has been shown [23, 36] that communication between an IoT device and its corresponding mobile App over the local network is typically unauthenticated and/or unencrypted. Additionally, consumer IoT devices expose themselves to the local network using SSDP and/or mDNS protocols that are typically used for local discovery. This feature enables attackers (or their malware hosted on a local device) to get in the middle of communication to eavesdrop or impersonate (i.e. man-in-the-middle can take full control over the device). These devices, with exposed local services, do not have any protection against flooding /denial-of-service attacks. We, therefore, deem devices that communicate over the local network to be at risk for this category of attacks, even though their MUD policies are well defined. This is because intended endpoints for local communications are not tightly defined. Moreover, all devices are vulnerable to ARP spoofing attack, as marked by light grey cells in Figure 6.4. We note that MUD specifications allow a local controller to restrict the local communications to a limited group – this is more conducive for enterprise networks.

Lastly, we see IoT devices with a large number of open ports which are not used (e.g. Telnet for HP Envy printer) [23], thus not specified in the MUD profile. These open ports make the device vulnerable, but having MUD policies in place would protect devices from being compromised by botnets such as Mirai, Bricker Bot, or Hajime that actively scan for open telnet ports and launch brute-force attack [40].

6.3.3   Monitoring MUD Activity

We saw (in Section 6.3.1) how MUD profiles can be translated into static and dynamic flow rules and applied at run-time on SDN switches to lock IoT traffic, thereby significantly reducing their attack surface.

Let us now focus on attacks that can be launched on IoT devices while still conforming to their MUD profiles. Specifically, we consider volumetric attacks that are not prevented by the MUD profile, since its ACLs simply allow or deny traffic, and there is no provision to limit rates. In this chapter, we show that a range of volumetric attacks (including ones directly on the IoT device and ones that reflect off the IoT device) are feasible in spite of MUD policy enforcement in the network. Fending off such attacks requires more sophisticated machinery that monitors the level of activity associated with each policy rule to detect anomalies. We leverage earlier studies [80] showing that IoT devices exhibit identifiable traffic patterns (with limited diversity of activity cycles and protocol use), making it feasible to develop machine learning methods for detecting abnormal behavior, which is otherwise difficult for general-purpose computers that exhibit much wider diversity in network behavior [12].

In order to identify sophisticated cyber-threats (volumetric attacks conforming to MUD profile), the activity of IoT flow rules specified by their MUD profile needs to be monitored and diagnosed. To do so, we need to periodically measure flow counters from the SDN switch, compute a set of attributes (features) for each device and stream them to their corresponding anomaly detectors.

We developed [39] a set of machine learning-based models to determine if an IoT device is involved in a volumetric attack or not, and if so, to determine the flow that contributes to the attack. Our models are trained with benign traffic profile of each device and detect attacks by flagging deviation from expected traffic pattern in device flows that are specified by the device MUD profile.

After generating flow rules for each IoT device (e.g. Figure 6.3 for the Canary camera), corresponding features of benign network activity are then extracted, using the counts of packets and bytes provided by each flow rule as features. This is because the size of the packet can vary for a given protocol.

Traffic features are also generated for multiple time-scales by retrieving flow counters (packet and byte counts) every minute and processing the counter values to generate values for the totals, means, and standard-deviations of packet and byte counts over sliding windows of two-, three- and four-minute as features, and including the original byte and packet count values as an additional two features, providing a total of 20 features for each flow rule at any point in time.

The anomaly detection models are based on the concept of one-class classification: device models are trained by features of benign traffic of their corresponding IoT device and are able to detect whether a traffic observation belongs to the trained class or not. Each anomaly detector uses a clustering-based outlier detection algorithm.

Lastly, in our recent work [82], we showed how IoT network operators can dynamically identify IoT devices using known MUD profiles and monitor their behavioral changes in their network. In that work, we described how the network behaviors of IoT devices are tracked at run-time, mapping their behavior to one of a set of known MUD profiles. This is needed for managing legacy IoTs that do not have support for the MUD standard. A MUD profile is a simple and environment-neutral description of IoT communications, and hence allows us to develop a simple model to identify corresponding devices. To do so, a behavioral profile is automatically generated and updated at run-time (in the form of a tree) for an IoT device, and a quantitative measure of its "similarity" to each of the known static MUD profiles (e.g. provided by manufacturers) is calculated. It is noted that computing similarity between two such profiles is a nontrivial task.

6.4 Conclusion

Vulnerable IoT devices are increasingly putting smart infrastructure at risk by exposing their networks unprotected to cyber-attackers. The IETF MUD standard aims to reduce the attack surface on IoTs by formally defining their expected network behavior. In this chapter, we have shown how behavioral modeling of IoT devices can prevent certain types of attacks while allowing for automatic detection of sophisticated volumetric attacks that are not prevented by the enforcement of MUD profiles. We discussed an SDN-based system empowered by machine learning models to continuously monitor the behavioral activity of MUD rules and detect anomalous patterns at fine-grained (per-flow) resolutions.

References

1 Lyu, M., Sherratt, D., Sivanathan, A. et al. (2017). Quantifying the reflective DDoS attack capability of household IoT devices. *Proceedings of the ACM WiSec*, Boston, Massachusetts, Jul 2017.

2 Sivanathan, A., Loi, F., Gharakheili, H.H., and Sivaraman, V. (2017). Experimental evaluation of cybersecurity threats to the smart-home. *Proceedings of IEEE ANTS*, pp. 1–6. Bhubaneswar, India, December 2017.

3 Sivaraman, V., Gharakheili, H.H., Fernandes, C. et al. (2018). Smart IoT devices in the home: security and privacy implications. *IEEE Technology and Society Magazine* 37 (2): 71–79.

4 Sivaraman, V., Chan, D., Earl, D., and Boreli, R. (2016). Smart-phones attacking smart-homes. *Proceedings of ACM WiSec*, pp. 195–200. Darmstadt, Germany, July 2016.

5 Sivaraman, V., Gharakheili, H.H., Vishwanath, A. et al. (2015). Network-level security and privacy control for smart-home IoT devices. *Proceedings of IEEE WiMob*, Abu Dhabi, UAE, October 2015.

6 Yu, T., Sekar, V., Seshan, S. et al. (2015). Handling a trillion (unfixable) flaws on a billion devices: rethinking network security for the internet-of-things. *Proceedings of ACM Workshop on HotNets*, Philadelphia, PA, USA, November 2015.

7 Lear, E., Romascanu, D., and Droms, R. (2019). Manufacturer Usage Description Specification. RFC 8520. IETF Secretariat, March 2019. https://www.rfc-editor.org/rfc/rfc8520.txt (accessed 23 September 2021).

8 Balaji, B., Bhattacharya, A., Fierro, G. et al. (2018). Brick: Metadata schema for portable smart building applications. *Applied Energy* 226: 1273–1292.

9 Paxson, V. (1999). Bro: A system for detecting network intruders in real-time. *Computer Networks* 31 (23–24): 2435–2463.

10 Roesch, M. (1999). Snort - lightweight intrusion detection for networks. *Proceedings of USENIX Conference on System Administration*, Seattle, Washington, November 1999.

11 Garcia-Teodoro, P., Diaz-Verdejo, J., Macia-Fernandez, G., and Vazquez, E. (2009). Anomaly-based network intrusion detection: techniques, systems and challenges. *Computer Security* 28 (1–2): 18–28.

12 Sommer, R. and Paxson, V. (2010). Outside the closed world: on using machine learning for network intrusion detection. *Proceedings of IEEE Security and Privacy (S&P)*, Berkeley, CA, USA, May 2010.

13 Lippmann, R., Haines, J.W., Fried, D.J. et al. (2000). The 1999 DARPA off-line intrusion detection evaluation. *Computer Networks* 34 (4): 579–595.

14 KDD Cup Data (1999). http://kdd.ics.uci.edu/databases/kddcup99/kddcup99 .html (accessed 23 September 2021).

15 Mahoney, M.V. and Chan, P.K. (2003). An analysis of the 1999 DARPA/Lincoln laboratory evaluation data for network anomaly detection. *International Symposium on Recent Advances in Intrusion Detection*, Pittsburgh, PA, USA, September 2003.

16 IoT Traffic Traces (2018). https://iotanalytics.unsw.edu.au/iottraces (accessed 23 September 2021).

17 Lim, S., Ha, J., Kim, H. et al. (2014). A SDN-oriented DDoS blocking scheme for botnet-based attacks. *Proceedings of ICUFN*, pp. 63–68. Shanghai, China, Jul 2014.

18 Niyaz, Q., Sun, W., and Javaid, A.Y. (2016). A Deep Learning Based DDoS Detection System in Software-Defined Networking (SDN). *arXiv preprint arXiv:1611.07400*.

19 Anderson, B. and McGrew, D. (2017). Machine learning for encrypted malware traffic classification: accounting for noisy labels and non-stationarity. *Proceedings of ACM SIGKDD*, pp. 1723–1732. Halifax, NS, Canada, August 2017.

20 Nokia Threat Intelligence Lab (2020). Threat Intelligence Report 2020. *Technical report CID210088*. NOKIA.

21 Ren, J., Dubois, D.J., Choffnes, D. et al. (2019). Information exposure from consumer IoT devices: a multidimensional, network-informed measurement approach. *Proceedings of ACM IMC*, Amsterdam, Netherlands, October 2019.

22 Andreeva, O., Gordeychik, S., Gritsai, G. et al. (2016). Industrial control systems and their online availability. *Kaspersky* 1–16.

23 Loi, F., Sivanathan, A., Gharakheili, H.H. et al. (2017). Systematically evaluating security and privacy for consumer IoT devices. *Proceedings of ACM IoT S&P*, Dallas, Texas, USA, November 2017.

24 Seals, T. (2020). Billions of Devices Open to Wi-Fi Eavesdropping Attacks. https://bit.ly/3a2JgRp (accessed 23 September 2021).

25 Apthorpe, N., Reisman, D., Sundaresan, S. et al. (2017). Spying on the smart home: privacy attacks and defenses on encrypted IoT traffic. *arXiv preprint arXiv:1708.05044*.

26 Acar, A., Fereidooni, H., Abera, T. et al. (2018). Peek-a-boo: I see your smart home activities, even encrypted! *arXiv preprint arXiv:1808.02741*.

27 Datta, T., Apthorpe, N., and Feamster, N. (2018). A developer-friendly library for smart home IoT privacy-preserving traffic obfuscation. *Proceedings of ACM IoT Security and Privacy*, Budapest, Hungary, August 2018.

28 Antonakakis, M., April, T., Bailey, M. et al. (2017). Understanding the Mirai botnet. *Proceedings of USENIX Security*, pp. 1093–1110.

29 Matherly, J. (2018). Shodan. https://www.shodan.io/ (accessed 23 September 2021).

30 Acar, G., Huang, D.Y., Li, F. et al. (2018). Web-based attacks to discover and control local IoT devices. *Proceedings of ACM IoT S&P*, Budapest, Hungary, August 2018.

31 Mehdi, S.A., Khalid, J., and Khayam, S.A. (2011). Revisiting traffic anomaly detection using software defined networking. *International Workshop on Recent Advances in Intrusion Detection*, pp. 161–180. Springer.

32 Odhaviya, R., Modi, A., Sheth, R., and Mathuria, A. (2017). Feasibility of idle port scanning using RST rate-limit. *Proceedings of ACM SIN*, pp. 224–228. Jaipur, India, October 2017.

33 Heberlein, L.T., Dias, G.V., Levitt, K.N. et al. (1989). A Network Security Monitor. *Technical report*. Davis, CA (USA): Lawrence Livermore National Lab., CA (USA) and California University.

34 Staniford-Chen, S., Cheung, S., Crawford, R. et al. (1996). GrIDS-a graph based intrusion detection system for large networks. *Proceedings of National Information Systems Security Conference*, Volume 1, pp. 361–370. Baltimore, MD.

35 Porras, P. and Valdes, A. (1998). Live traffic analysis of TCP/IP gateways. *Proceedings of NDSS*, San Diego, CA.

36 Demetriou, S., Zhang, N., Lee, Y. et al. (2017). HanGuard: SDN-driven protection of smart home WiFi devices from malicious mobile apps. *Proceedings of ACM WiSec*, Boston, Massachusetts, July 2017.

37 Hamza, A., Gharkheili, H.H., and Sivaraman, V. (2018). Combining MUD policies with SDN for IoT intrusion detection. *Proceedings of ACM workshop on IoT S&P*, Budapest, Hungary, August 2018.

38 Boddy, S. and J. Shattuck (2017). The Hunt for IoT: The Rise of Thingbots. *Technical report*. F5 Labs. https://bit.ly/3jvBHcO.

39 Hamza, A., Gharakheili, H.H., Benson, T.A., and Sivaraman, V. (2019). Detecting volumetric attacks on IoT devices via SDN-based monitoring of MUD activity. *Proceedings of ACM SOSR*, San Jose, CA, USA, April 2019.

40 Cisco Systems Inc. (2017). Midyear Cybersecurity Report. *Technical report*. Cisco Technology News Site, San Jose, CA, USA.

41 Cisco (2018). Cisco 2018 Annual Cybersecurity Report. *Technical report*. Cisco Technology News Site, San Jose, CA, USA.

42 Micro, T. (2017). Persirai: New Internet of Things (IoT) Botnet Targets IP Cameras. https://bit.ly/3b4PJLL (accessed 23 September 2021).

43 Micro, T. (2019). Bashlite IoT Malware Updated with Mining and Backdoor Commands, Targets WeMo Devices. https://bit.ly/2xJGH8E (accessed 23 September 2021).

44 Smith (2018). Mirai Okiru: New DDoS botnet targets ARC-based IoT devices. https://bit.ly/2Ui3csX (accessed 23 September 2021).

45 Seals, T. (2018). Wicked Botnet Uses Passel of Exploits to Target IoT. https://bit.ly/33r7m5S (accessed 23 September 2021).

46 Avast (2018). Torii botnet - Not another Mirai variant. https://bit.ly/2UfOzXg (accessed 23 September 2021).

47 Lakhina, A., Crovella, M., and Diot, C. (2005). Mining anomalies using traffic feature distributions. *ACM SIGCOMM Computer Communication Review*, Volume 35, pp. 217–228. ACM.

48 Kumar, K., Joshi, R.C., and Singh, K. (2007). A distributed approach using entropy to detect DDoS attacks in ISP domain. *Proceedings of IEEE ICSPCN*.

49 Giotis, K., Argyropoulos, C., Androulidakis, G. et al. (2014). Combining Open-Flow and sFlow for an effective and scalable anomaly detection and mitigation mechanism on SDN environments. *Computer Networks* 62: 122–136.

50 Braga, R., Mota, E., and Passito, A. (2010). Lightweight DDoS flooding attack detection using NOX/OpenFlow. *Proceedings of IEEE LCN*, pp. 408–415. Denver, CO, USA, October 2010.

51 Cui, Y., Yan, L., Li, S. et al. (2016). SD-Anti-DDoS: Fast and efficient DDoS defense in software-defined networks. *Journal of Network and Computer Applications* 68: 65–79.

52 Tang, T.A., Mhamdi, L., McLernon, D. et al. (2016). Deep learning approach for network intrusion detection in software defined networking. *Proceedings of IEEE WINCOM*, Fez, Morocco, October 2016.

53 Bhunia, S.S. and Gurusamy, M. (2017). Dynamic attack detection and mitigation in IoT using SDN. *Proceedings of ITNAC*, pp. 1–6. IEEE.

54 Huang, D.Y., Apthorpe, N., Acar, G. et al. (2019). IoT Inspector: Crowdsourcing Labeled Network Traffic from Smart Home Devices at Scale. *arXiv preprint arXiv:1909.09848*.

55 Jinhua, G. and Kejian, X. (2013). ARP spoofing detection algorithm using ICMP protocol. *Proceedings of IEEE ICCI*, pp. 1–6. Coimbatore, India, January 2013.

56 Bharti, A.K., Goyal, M., and Chaudhary, M. (2013). A review on detection of session hijacking and IP spoofing. *International Journal of Advanced Research in Computer Science* 4 (9). 104–107.

57 Garofalo, A., Di Sarno, C., Coppolino, L., and DÁntonio, S. (2013). A GPS spoofing resilient WAMS for smart grid. *European Workshop on Dependable Computing*, pp. 134–147. Springer.

58 Ramesh, G., Krishnamurthi, I., and Kumar, K.S.S. (2014). An efficacious method for detecting phishing webpages through target domain identification. *Decision Support Systems* 61: 12–22.

59 Ateniese, G. and Mangard, S. (2001). A new approach to DNS security (DNSSEC). *Proceedings of ACM CCS*, Philadelphia, PA, USA, November 2001.

60 dos Santos, D. (2019). Sabotaging Common IoT Devices in Smart Buildings by Exploiting Unencrypted Protocols. https://bit.ly/2IZ6TyG (accessed 23 September 2021).

61 Montalbano, E. (2020). Hacker Leaks More Than 500K Telnet Credentials for IoT Devices. https://bit.ly/33ukHdR (accessed 23 September 2021).

62 Kan, M. (2016). IoT botnet highlights the dangers of default passwords. https://bit.ly/2UfVCPD (accessed 23 September 2021).

63 Feng, Y., Wang, W., Weng, Y., and Zhang, H. (2017). A replay-attack resistant authentication scheme for the internet of things. *Proceedings of IEEE CSE and IEEE EUC*, Guangzhou, China, August 2017.

64 Wynn, M., Tillotson, K., Kao, R. et al. (2017). Sexual intimacy in the age of smart devices: are we practicing safe IoT? *Proceedings of ACM IoT S&P*, Dallas, Texas, USA, November 2017.

65 SSLLabs (2020). SSL and TLS Deployment Best Practices. https://bit.ly/392rnko (accessed 23 September 2021).

66 Giani, A., Berk, V.H., and Cybenko, G.V. (2006). Data exfiltration and covert channels. *Sensors, and Command, Control, Communications, and Intelligence (C3I) Technologies for Homeland Security and Homeland Defense V*, Volume 6201, pp. 5–15. International Society for Optics and Photonics.

67 Mitre (2018). Exfiltration. https://attack.mitre.org/tactics/TA0036/ (accessed 23 September 2021).

68 Do, Q., Martini, B., and Choo, K.-K.R. (2016). A data exfiltration and remote exploitation attack on consumer 3D printers. *IEEE Transactions on Information Forensics and Security* 11 (10): 2174–2186.

69 Yaqoob, I., Ahmed, E., ur Rehman, M.H. et al. (2017). The rise of ransomware and emerging security challenges in the internet of things. *Computer Networks* 129: 444–458.

70 Ronen, E. and Shamir, A. (2016). Extended functionality attacks on IoT devices: the case of smart lights. *Proceedings of IEEE EuroS&P*, Saarbrucken, Germany, May 2016.

71 Ahmed, J., Gharakheili, H.H., Raza, Q. et al. (2019). Monitoring enterprise DNS queries for detecting data exfiltration from internal hosts. *IEEE Transactions on Network and Service Management* 17 (1): 265–279.

72 IFAC (2019). Jamming attacks: a major threat to controlling over wireless channels. https://bit.ly/2wljyJg (accessed 23 September 2021).

73 Tang, X., Ren, P., and Han, Z. (2018). Jamming mitigation via hierarchical security game for IoT communications. *IEEE Access* 6: 5766–5779.

74 Hamza, A., Gharkheili, H.H., Pering, T., and Sivaraman, V. (2020). Combining device behavioral models and building schema for cyber-security of large-scale IoT infrastructure. *Under Review*.

75 Pirretti, M., Zhu, S., Vijaykrishnan, N. et al. (2006). The sleep deprivation attack in sensor networks: analysis and methods of defense. *International Journal of Distributed Sensor Networks* 2 (3): 267–287.

76 Nash, D.C., Martin, T.L., Ha, D.S., and Hsiao, M.S. (2005). Towards an intrusion detection system for battery exhaustion attacks on mobile computing devices. Proceedings of IEEE International Conference on Pervasive Computing and Communications Workshops, Kauai Island, HI, USA, March 2005.

77 Hristozov, S., Huber, M., and Sigl, G. (2019). Protecting RESTful IoT Devices from Battery Exhaustion DoS Attacks. *arXiv preprint arXiv:1911.08134*.

78 Surbatovich, M., Aljuraidan, J., Bauer, L. et al. (2017). Some recipes can do more than spoil your appetite: analyzing the security and privacy risks of

IFTTT recipes. *Proceedings of WWW*, pp. 1501–1510. Perth, Australia, April 2017.

79 Sapountzis, N., Sun, R., and Oliveira, D. (2018). DDIFT: Decentralized dynamic information flow tracking for IoT privacy and security. *Workshop on Decentralized IoT Systems and Security (DISS)*, Perth Australia, April 2018.

80 Sivanathan, A., Gharakheili, H.H., and Sivaraman, V. (2020). Managing IoT cyber-security using programmable telemetry and machine learning. *IEEE Transactions on Network and Service Management* 17 (1): 60–74.

81 Sivanathan, A., D. Sherratt, H. Habibi Gharakheili, A. Radford, et al. Characterizing and classifying IoT traffic in smart cities and campuses. *Proceedings of IEEE INFOCOM workshop on SmartCity*, Atlanta, Georgia, USA, May 2017.

82 Hamza, A., Ranathunga, D., Gharakheili, H.H. et al. (2020). Verifying and monitoring IoTs network behavior using MUD profiles. *IEEE Transactions on Dependable and Secure Computing*.

83 UNSW MUD repository (2018). https://iotanalytics.unsw.edu.au/mudprofiles (accessed 23 September 2021).

84 Insecam (2018). Network live IP video cameras directory. http://www.insecam .org (accessed 23 September 2021).

7

Integrity of IoT Network Flow Records in Encrypted Traffic Analytics

Aswani Kumar Cherukuri[1], Ikram Sumaiya Thaseen[1], Gang Li[2], Xiao Liu[2], Vinamra Das[1], and Aditya Raj[1]

[1] School of Information Technology and Engineering, Vellore Institute of Technology, Vellore, Tamil Nadu, India
[2] School of Information Technology, Deakin University, Geelong, Melbourne, Australia

7.1 Introduction

Traffic behavior analysis techniques work with encrypted protocols, as they are not dependent on the packet payload. Monitoring network traffic can identify the data conversed from the sender to distinct clients and extensive research has been performed on attack detection by analyzing the contents of the message. It is a known fact that IoT devices are easier to infiltrate [1, 2], and new stories arise of how IoT devices are compromised and large-scale attacks are launched on those devices. Hence, securing the IoT environment is a challenge for network operators in large organizations [3–5]. These operators scale up these devices online. There are major security breaches in advanced technologies such as smart homes, smart industries [6, 7]. Hence, analyzing the IoT network traffic to identify potential security breaches, malicious activities is highly essential. Large-scale attacks have been unveiled by negotiating and utilizing many IoT devices [8]. Hence, it is important to analyze the IoT traffic. The most compelling reason to analyze IoT traffic is to enhance cyber-security. Security gaps in large organizations [9] are due to unauthorized access which constitutes up to 65% of information security breaches according to the survey conducted in 2015. Most of the IoT traffic is unencrypted which clearly means that attackers can execute man-in-the-middle (MiTM) attacks. Attackers can get in between devices to steal or alter the data by easily accessing the unencrypted stream of data. This has led researchers to develop network security mechanisms to identify and mitigate attacks in the IoT environment. Several network-level mechanisms to analyze

Security and Privacy in the Internet of Things: Architectures, Techniques, and Applications,
First Edition. Edited by Ali Ismail Awad and Jemal Abawajy.

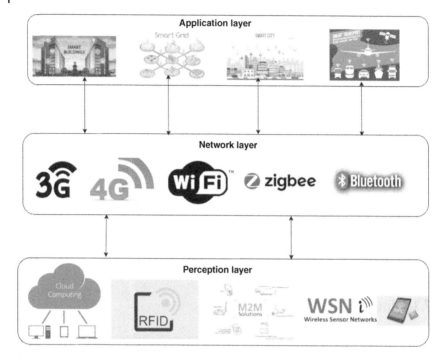

Figure 7.1   Three-layered IoT architecture. Source: icon0/Adobe Stock; Unknown author/Wikimedia Commons; Connectivity Standards Alliance/Wikimedia Commons; Wireless Sensor Networks.

traffic and identify attacks are recommended by researchers considering the huge heterogeneity in IoT devices [10, 11].

Figure 7.1 shows a three-layered IoT architecture [7]. The bottom layer, the perception layer, is comprehensively used in RFID, GPS, biometrics, sensors, EPC, barcode technology, and other advanced technologies to acquisition data in a timely manner. A large number of Internet of Things devices are connected to the network layer which supports its own equipment, such that data can be uploaded and received through Internet. The virtual data information is processed, classified, and finally transmitted to the application layer. Thus, the application of the IoT technology in the entire architecture system is analyzed. Traffic created by IoT devices vary considerably. This traffic will depend on whether or not an edge processor is utilized, how much intelligence is present, and the applications which are running on the edge processor. IoT traffic characteristics include continuous traffic flow, endpoints to collect data, periodic data with automatic delivery and interactive communications with actuators if they are present.

Traffic gathering can happen using IPF IX or NetFlow at the flow level or packet level [12, 13]. Passive observers such as ISPs periodically collect flow-level traffic for network administration. There are several forms of passive monitoring. Manual assessment is performed for simple monitoring as the amount of analyzed data is small. If the data to be monitored is huge, it is technologically demanding to handle and save the data. Network traffic is detected from an observation point. Packet capturing can be manual or automatic [14]. Malware behavior monitoring and analysis is performed by the automated technique [15]. Insider attacks are recognized by many existing studies which are mainly based on two approaches: packet inspection and flow-based technique, both of which can be utilized in IDS, network forensics tools, and Security Incident and Event Management (SIEM) systems [15–17].

Transmission of flow data is done in various forms, for example NetFlow [18] built by CISCO, SFlow [17], JFlow [19], and IPFIX [20] developed by IETF. Varieties of tasks are executed by the abovementioned applications such as monitoring of traffic, detecting unauthorized activity, and tracing the DoS attack source. This is typically accomplished by analyzing the flow of current traffic and detecting any irregularity in comparison to the past traffic profile. Flow observation [21] has several advantages:

i. it is quicker than other techniques using the same hardware;
ii. the privacy concern is less because the payload [16] is not processed in comparison to packet capture or deep packet inspection (DPI);
iii. timely and quick analysis for huge networks; and
iv. effective in monitoring network usage, discovering scans, and intrusions in network and the spread of malware.

With these advantages, the system can analyze the information collected from all IoT devices connected to the network and can provide security.

Sensitive user communications can be revealed even if IoT devices send the network traffic in encrypted mode. Metadata such as send/receive rates, IP, and TCP packet headers are accessible to the adversary [10]. ISPs frequently collect the metadata for traffic study. The network viewers are prevented from analyzing the sensitive information in traffic [22] due to the increase in pervasiveness of encryption. It is however noted that in spite of encryption at the transport layer [23], passive adversary can deduce sensitive events from metadata of smart home traffic [24]. Hence, it is essential to analyze and secure the meta data from these devices even when encryption is implemented in the network. It is therefore possible for an adversary to acquire and analyze IoT devices. An attacker can collect traffic data to detect sensitive events in end-user traffic [3].

Traffic flows are analyzed to understand and infer the details of the network traffic. When the traffic is encrypted, network flow level data is analyzed to understand any malicious activity under the encrypted communication. Flow data is the first source of information which analyzes malicious activity in encrypted communication. Hence, protecting the integrity of flow data is an important challenge. Therefore, in this chapter, we propose a technique for protecting the integrity of the flow-level data. Main contribution of our work is as follows: We study the flow-based approaches available in the literature to analyze the encrypted network traffic. In particular, we present the need to protect the integrity of this telemetry. Then we propose hash-based message authentication codes to protect the integrity of the network telemetry. To the best of our knowledge, this is the first proposal in the literature toward this direction. Rest of the chapter is organized as follows: The background of the encrypted traffic analytics is discussed in Section 7.2. Flow-based telemetry and threats to telemetry data is outlined in Section 7.3. An authentication technique using hashing for flow-based data is discussed in Section 7.4. A case study is demonstrated with three different cases in Section 7.5 to simulate the various cases proposed for authenticating the flow-based data using benchmark IoT flow-based dataset. Section 7.6 concludes the chapter.

7.2 Background

Flow data is considered as a single packet flow in a network comprising of protocol, IP, and ports of source and destination. Packets are accumulated into flow records that aggregate the amount of transferred data, the total number of packets, and other information captured from the network and transport layer. Any payload can be encrypted for preserving privacy and thereby effectively securing the communication. There is no difference among authorized and unlawful communication from an encryption point of view. There is a challenge in network monitoring due to the variety and dynamism of today's networks [4]. Packet capture solutions work well in specific cases, but they cannot manage with the scalability, flexibility, and ease-of-use of flow data. There is a need for flow-based analysis due to the increase in network speeds, discernibility gaps initiated by migration to cloud, IoT, and software-defined networking [25].

7.2.1 Encrypted Traffic Analytics (ETA)

Encrypted traffic is analyzed by payload- and feature-based classification techniques, which are surveyed widely in the literature [26] and characterized using a conventional nomenclature. A complete assessment of the feature classification techniques was performed and presented with their flaws and assets. Since traditional payload-based approaches are incapable of handling encrypted traffic, many machine-learning techniques are proposed [27]. A hybrid method was developed [28] combining signature and statistical techniques to address the concern. In the initial phase, signature matching techniques were utilized for SSL/TLS traffic. In addition, statistical methods are applied to identify type of application. The approach was proven to detect SSL/TLS traffic above 99% accuracy and with 94.52% in F-score for identifying the protocols. Montigny–Leboeuf established a number of indicators that reveal necessary communication dynamics according to the information collected from monitoring traffic flow packet headers [29]. Anderson and David [30] have developed an approach that contains flow metadata, namely the unencrypted TLS handshake data and pointers to related flows. The authors used a diverse approach in comparison to other flow-based techniques because it not only uses the metadata but also utilizes detailed information about the flow and related flows. Common attributes were mined [31] from the size and timing information of packets which are packet volume integrated with forward or reverse direction and interarrival time. These attributes differentiate between various categories of network traffic.

Encrypted traffic without analyzing the payload, port number, and IP addresses were evaluated [32] by deploying three machine learning approaches. The experimental study inferred that in comparison to K-means clustering technique and C4.5, superior performance resulted in multiobjective genetic algorithm (MOGA). Five learning algorithms were evaluated by authors [33] using flow-based features, without considering the source/destination ports, IP addresses, and payload. Different machine learning approaches were used including Ripper, Naive Bayes, C4.5, SVM, and Ada Boost. The analysis showed that C4.5 performed better on the categorization of SSH and Skype traffic on diverse networks. Alshammari and Nur [34] have implemented a machine learning-based technique without analyzing the payload, ports, and IP but using the packet header attributes and statistical flow attributes to disclose encrypted application channels in network traffic with high precision. Anomaly recognition for encrypted web accesses [35] is implemented without decryption by utilizing time and data volume. Any private information or massive preoperation is not required as in regular encrypted traffic study. Various attacks were recognized with high accuracy.

7.2.2 Techniques for ETA

For encrypted traffic analysis, different techniques are recommended and used in practice including packet sampling [36], SNMP [37], Wi-Fi packet sniffing [38, 39], and flow-based telemetry [40, 41]. Contemporary telemetry techniques can be characterized into packet-based [36, 42, 43] or flow-based [40, 41, 44]. A method regularly used is sFlow [36] which samples packets obtained from the network switches in a random manner. The packets of sFlow were captured from elephant flows (long duration and dense traffic). Due to this, mice flows can be missed resulting in an inaccurate analysis. Everflow [42] addresses this issue by analyzing packets of TCP, FIN, SYN, and RST gathered at data center switches using the match and mirror concept. The throughput of flow was evaluated by Planck [43] by mirroring multiple port traffic after which a high-rate sampling is executed by the collector. Thus, only limited visibility into network traffic flows is provided by packet-level telemetry.

In literature, few researchers have analyzed the encrypted traffic and proposed solutions. A mechanism with 23 traffic features of 276 dimensions for 12 packets was deployed [12] which resulted in excessive computational cost. Maiti et al. [13] have presented a mechanism which consists of 30,000 frames with 49 traffic features of long duration traffic required by the detect devices. A variable size time window is utilized [6] to analyze the IoT devices communication rate. Yet, those techniques [6, 12, 13] require huge memory to store the attributes, computation cost is high, and real time detection is challenging [45, 46]. An attacker can recognize sensitive data such as type and specific behavior of IoT traffic from the IoT device [47]. Machine learning techniques [46] with contributing features of encrypted IoT traffic are trained to avoid thumbprint attack.

7.2.3 Hashing for Flow Record Authentication

Many IoT devices communicate with the cloud to store and process the information. With the initiation of long range of telemetry flow record [48], and the related increase in the communication range, there is an increase in the risk of flow record being negotiated. For instance, a replay attack can be initiated in which a flow record or a piece of flow record can be seized and then utilized at a later duration maliciously [49]. Hence, improved systems are required for telemetry. Hence, securing telemetry flow data is very important. Methods have to be developed to verify that the integrity of the flow record is not compromised and the session is authorized. A hash algorithm is considered secure because it is computationally infeasible to determine the message from a given message digest or hash value. It is also not possible to determine two distinct messages which result in the same message digest. Due to these properties of hash functions, if the message integrity is compromised, it results in a detectable change in the hash value. However, a modified message will produce a detectable change in the hash value. A variety of hash algorithms can be deployed to generate a hash value, and subsequently, a message authentication code. MD2, MD4, and MD5 are additional examples of hash algorithms. Each of these algorithms produces a 128-bit length of message digest. To prevent a replay attack and preserve the integrity of the telemetry, we need to follow this sequence: the flow record originating device requests a code from the flow record receiving device. The receiving device randomly selects a code and transfers the code to the sending device. The sender sends the flow record along with the generated hash, key, and code. Thus, with this sequence of events, an unreceptive sender replaying an earlier code cannot fool the receiver.

7.3 Flow Based Telemetry

7.3.1 Flow Metadata

In order to detect the anomaly in the network traffic, network traffic monitoring in a passive mode is essential. While monitoring the traffic, anomaly detection systems check for any deviations from the normal traffic behavior. While monitoring, encrypted network traffic, anomaly detection systems depends on flow telemetry. The metadata about network information is traditionally captured using a flow-monitoring system which is exported to the receiver by commercial switches tailored with Netflow [41] or IPFIX [50]. This data are important because traffic is encrypted and DPI is not suitable. But there are two weaknesses: (i) a flow record is exported after it expires which specifies that it is not real time, and (ii) maintaining and updating flow records in the switch [45] can be computationally cost ineffective. The limitations of Netflow are overcome by FlowRadar [40] which implements an encoded hash table and regularly exporting flows (e.g. 10 ms) with low memory overheads. However, commercial switches existing on the market still did not support FlowRadar. SDN APIs [44] are used for implementing a flow-level telemetry which can analyze traffic flows at the minimum cost.

7.3.2 Flow-Based Approaches

Flow data has been widely used to perform malware or intrusion detection [51–54]. BotFinder [55] is a major correlation approach that utilizes only NetFlow features. Clusters are identified using unsupervised machine learning approaches [56] that represent malware communications gathered from malware sandboxes. As more data attributes are utilized, the robust approach is not episodic in Net-Flow [57] and also works against malwares that do not possess communication characteristics. These techniques also work when the malware is only detected for a short time interval. This is considered to be a major asset, since labeled, huge malware datasets are often restricted by sandbox resources. Another technique is BotHunter [51]. In this approach, the communication between external objects and internal assets is tracked and a trail evidence of communication events is developed that is verified with state-based infection sequence prototype. BotHunter is implemented by integrating signatures obtained from Snort and anomaly detection. BotSniffer [58] analyses similarity patterns and spatial-temporal correlations to characterize the botnet command and control traffic. BotMiner [52] is another approach which cannot identify single infected host with high efficiency.

Meidan et al. [59] classified a single TCP flow on the network by utilizing machine learning techniques. Nearly, 300 features of packet level and flow level are employed. Contributing features are minimum, median, Time-To-Live (TTL), total packets, total packet count with reset (RST) flag, total bytes sent and received

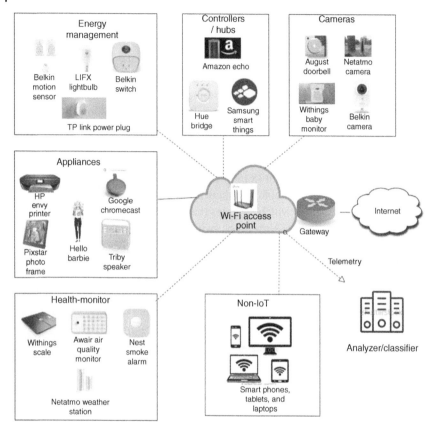

Figure 7.2 Test bed architecture of different IoT and non-IoT devices telemetry collected across the structure and fed to the classifier. Source: Belkin International, Inc.; LIFX; TP-Link Corporation Limited; Amazon, Inc.; Signify Holding; Samsung; August Home; Netatmo; Withings; HP Development Company, L.P.; Pix-Star; photology1971/Adobe stock; Invoxia; Google, Inc.; Awair Inc.

ratio, and the Alexa rank of the server. Figure 7.2 shows test bed architecture of various IoT, non-IoT devices telemetry fed to the classifier for decision-making. In the diagram, a Wi-Fi Access point is considered as the central node integrated with the Internet gateway. There are a total of 20 distinctive IoT devices signifying various classes along with different non-IoT devices. In the illustration, specific Internet-linked devices (e.g. smoke sensors and camera) are referred as IoT devices and general-purpose devices such as phones and laptops are in the non-IoT category. Telemetry is gathered across the network which is fed to the classifier. Flow telemetry of these IoT devices is collected for further analysis by the classifier. The role of classifier is to analyze the flow telemetry data to check the anomalies in the traffic.

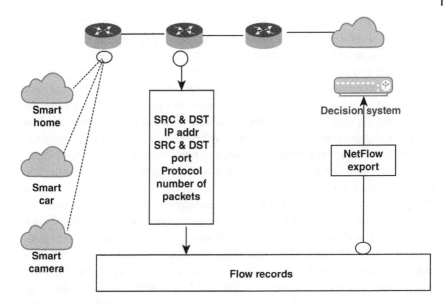

Figure 7.3 IoT-based telemetry.

Figure 7.3 shows an illustration of IoT-based telemetry. For representative purpose, any IoT device at Smart Home, Smart Car, and Smart Camera can communicate through gateway to the cloud server where all the information is stored.
There can be other IoT devices also linked to the gateway. The gateways which are
also known as the collector gather the metadata from the network traffic namely
the flow-related information and export it to NetFlow which in turn sends to the
decision system to decide on the traffic pattern. As discussed above, these flow
records are sent to the classifier to verify any anomalies in the traffic. Considering
the encrypted traffic environments, the role of analyzer is much more important.
While analyzing the encrypted traffic, classifiers would focus on the flow telemetry
data as they are generally not encrypted. From this telemetry, classifiers identify
the anomaly or intrusion in the traffic. Attackers can try to compromise this unencrypted flow telemetry. Hence, in this chapter, focus is on integrity of this telemetry
data collected by the gateways and propose methods to ensure integrity of this data
before they are analyzed for detecting patterns.

7.3.3 Threats on Flow Telemetry

It is possible for an attacker to analyze, insert, or modify packets which will
be exported to the receiver. Hence, IP Flow data can be observed or forged,
thereby NetFlow Collector can be a target for future attacks [55]. The flow records
which are unencrypted provide the information regarding traffic behavior,

communication endpoints, and active flows in the network to the attacker. With the collected information, it is likely that the user pattern can be eavesdropped and can plan to hide future attacks [5]. A flow record with the IP of source and destination may disclose sensitive data about the end-user actions; however, with only the source and destination IP network values, a flow record may not provide enough information to the attacker. Flow records deployed for security and/or accounting applications are usually vulnerable to attacks by falsifying the flow records which are exported. For instance to prevent the identification of an attack, or to deceive the service provider, attacker may tamper the flow data. This is implemented by inserting bogus flow records that act like original or modifying the flow records on the route between sender and the receiver. Denial-of-Service attack is also possible on the Net Flow collector by consuming maximum resources from the server such that it is not possible to decode or capture many Net Flow Export packets. Few authors have proposed solutions to overcome threats on TLS-aware telemetry [31]. The authors analyzed the desirable characteristics of intraflow data attributes and deployed machine learning classifiers. From the literature, it is evident that with the absence of payload, it is feasible to capture data from the metadata of packets. Hence, a flow record-based authentication is required between the Exporter and the Collector to preserve the integrity of the flow records in the IoT domain. This is achieved by proposing hash-based approaches. Details of the proposal and demonstrations are given in the following sections.

7.4 Hashing-Based MAC for Telemetry Data

In order to address the problems discussed above, different authentication mechanisms are proposed in this section. Every mechanism assumes that the flow record data originates from an IoT device and sends to an intermediate gateway. Another gateway sends the collected flow record to the NetFlow Export which is assumed as the new client. The client implements the authentication mechanism and the NetFlow Export is the server which verifies whether the authentication succeeded or failed for each flow record. Secure Hash Algorithm is used for hash calculation. Following are the different cases of the proposal.

Case 1: Hashing-Based Flow Telemetry Authentication

Figure 7.4 illustrates the implementation of a hash function with no encryption for flow record authentication. In this case, the two end parties namely the gateway and NetFlow Export share a common secret value S. The gateway computes the hash value after appending the flow record along with the secret value. The calculated hash value of the flow record is communicated along with the flow record to the NetFlow Export. At the NetFlow Export, the secret value

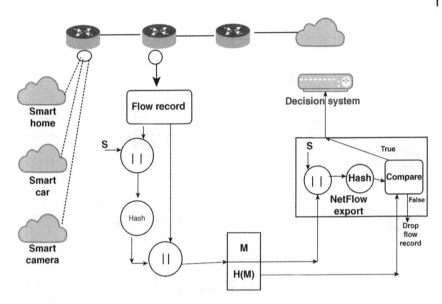

Figure 7.4 Flow record authentication using end-to-end hashing.

S is reused to calculate the new hash of the received flow record, a comparison of the received hash value from the gateway and calculated hash value is performed. The flow record is authenticated only if the hash value matches and sent to decision system. Otherwise, the flow record is discarded.

Case 2: Symmetric Encryption-Based Flow Telemetry Authentication

Figure 7.5 illustrates the authentication of flow record using hashing with symmetric encryption. At the gateway, a hash of the flow record is calculated and encrypted using symmetric encryption. A variety of symmetric encryption techniques are available. Examples include DES, Triple DES, AES, IDEA, Blowfish, and CAST. The proposed case is implemented using AES in Cipher Block Chaining mode. The encrypted flow record is sent to the NetFlow Export. The gateway and NetFlow share the same secret key. At the NetFlow, the flow record is decrypted and a new hash of the flow record is computed. The received hash is then compared with new hash to check whether the flow record sent from the gateway is not altered. If both the hash value matches, the process is same as is in the previous case. Authentication is achieved using the hashing technique. Confidentiality is also accomplished as encryption is applied to the entire flow record along with the hash code.

Case 3: Asymmetric Encryption-Based Flow Telemetry Authentication

Figure 7.6 shows an authentication mechanism using hashing with asymmetric encryption. In this case, only the hash value is encrypted, using public key encryption and using the gateway's private key. Authentication is achieved in

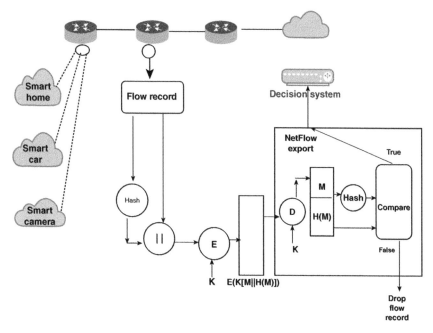

Figure 7.5 Flow record authentication using end-to-end hashing with symmetric encryption.

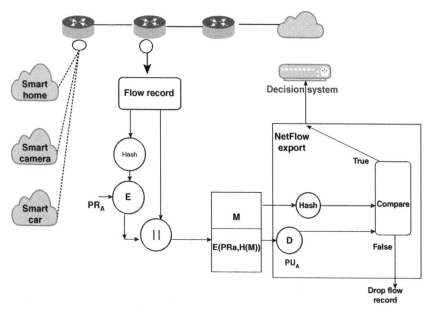

Figure 7.6 Flow record authentication using end-to-end hashing with asymmetric encryption.

this scenario. A digital signature is also provided because only the gateway could have created the encrypted hash value. At the NetFlow Export, the public key of the gateway is used to decrypt the encrypted hash value. A new hash digest is computed from the received flow record. Then, both the hash values are compared to verify the integrity of the flow record. Fake flow records are discarded and authenticated flow records are sent to decision system for classifying the flow records.

7.5 Experimental Analysis

In order to validate the proposed models, experiments are conducted in a simulated environment with help of Python. A client server communication is established by creating a socket and sending every flow record from client to server. Three different scenarios are implemented: (i) hashed flow record; (ii) symmetric with hash flow record and (iii) asymmetric with hash flow record. For every case, we assume that the gateway is the client and the NetFlow Export is the server. We have simulated a MITM attacker and shown that how the attack will fail because of the flow record authentication.

7.5.1 Hashed Flow Records

The flow record data which are sent from client are source address, destination address, source port, destination port, and protocol. In the server side, every flow record is obtained, and new hash value is computed. The original hash value sent from the client is compared with the new calculated hash value to verify the flow record integrity. Figure 7.7 shows the client server connection establishment which is successful from the server side. A unique id is generated for every flow record sent to the server. The flow record details are stored in a file and retrieved for analysis. Figure 7.8 illustrates the notification to the client that the server has accepted the client connection and the flow record. Figure 7.9 shows the server side computation of the hash for the flow record and verification of hash. An authenticated response indicates that the hash verification is successful after comparison of generated and received hash. Figure 7.10 shows that the MITM attacker failed at the server side. This is because the hash value of the original flow record computed at the authenticated client is sent to the server. When a forged flow record is sent from the attacker's side, a new hash value will be generated which will mismatch with the received hash value from the authenticated client. Figure 7.11 is the illustration of the failed authentication notified to the fraud client.

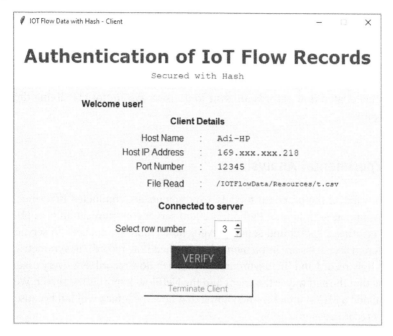

Figure 7.7　Client server establishment to transfer flow records. Source: The PHP Group.

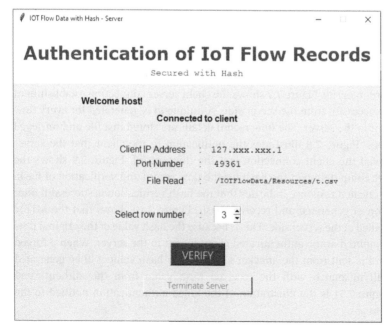

Figure 7.8　Server connected to clients using hash only approach. Source: The PHP Group.

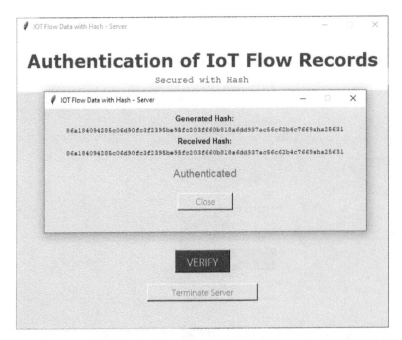

Figure 7.9 Server side hash computation of IoT flow records. Source: The PHP Group.

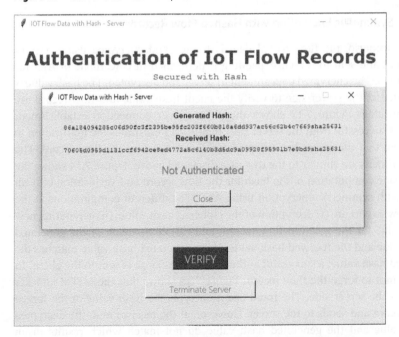

Figure 7.10 MITM attack failed at the server side. Source: The PHP Group.

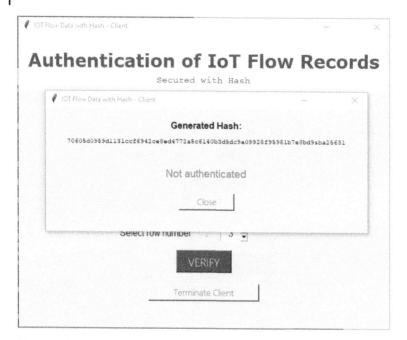

Figure 7.11 MITM attack response to the client. Source: The PHP Group.

7.5.2 Symmetric Encryption with Hashed Flow Records

In this scenario, the flow record data is hashed and encrypted using AES in the client and sent to server. A padding is performed in CBC mode of AES for encryption. The encrypted data are sent to server and decrypted. The hash value is computed at the server side to verify the client hash value and flow record are authenticated. Figure 7.12 shows the client server connection establishment which is successful from the server side. Figure 7.13 illustrates the notification to the client that the server has accepted the connection and the flow record. The first two steps are similar to the first case discussed above. Figure 7.14 shows the server side computation of the hash for the flow record and verification of hash along with symmetric encryption using AES. The different computations at the NetFlow Export are (i) decryption of the ciphered hash value; (ii) generating new hash value for the received flow record; and (iii) comparison of the generated hash value and the received hash value. If the generated hash value matches the received hash value, a successful authentication response is sent to the client. In an attempt to forge the flow record, Figure 7.15 shows that the MITM attacker failed at the server side. The fraud client encrypts the hash value of the forged flow record and sends to the server. However, at the receiver end, the decrypted hash value and the generated hash value do not match which results in an

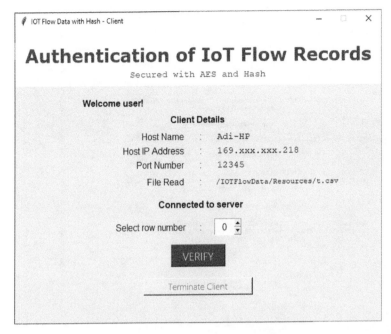

Figure 7.12 Client server establishment using symmetric hash. Source: The PHP Group.

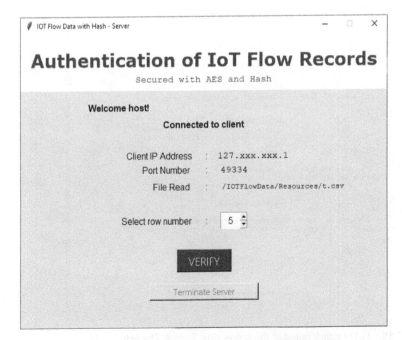

Figure 7.13 Server connected to the client. Source: The PHP Group.

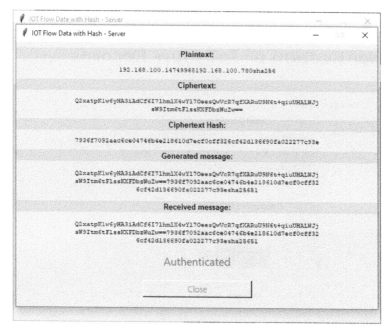

Figure 7.14 Server validating the client using AES with Hash. Source: The PHP Group.

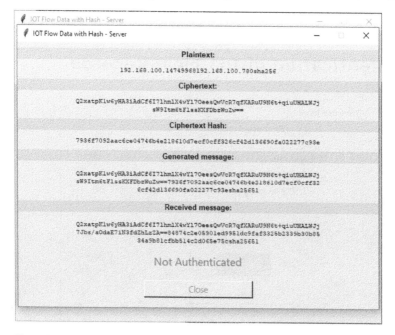

Figure 7.15 MITM attack failed at the server side. Source: The PHP Group.

Figure 7.16 MITM attack responded back to the client. Source: The PHP Group.

authentication failure. Figure 7.16 is the illustration of the failed authentication notified to the client.

7.5.3 Asymmetric Encryption with Hashed Flow Records

The flow record is hashed and encrypted using RSA in the client and sent to server. Similar to the symmetric approach, a hash value is computed using SHA algorithm on the flow record and then encrypted using RSA from the client and sent to the server. Whenever, a trusted client sends a flow record it is authenticated at the server side. However, when a fraudulent client sends a modified flow record, there is a mismatch in the hash value computation at the server side and hence authentication fails at the server side. Figure 7.17 shows the client server connection establishment which is successful from the server side. Figure 7.18 illustrates the notification to the client that the server has accepted the connection and the flow record. Figure 7.19 shows the server side computation of the hash for the flow record and verification of hash along with asymmetric encryption using RSA. Similar to Case 2, various computations are performed at the NetFlow Export. Figure 7.20 shows the client receiving the authenticated message from the server. Figure 7.21 shows the MITM attacker failed at the server side. Figure 7.22 is the illustration of the failed authentication notified to the client.

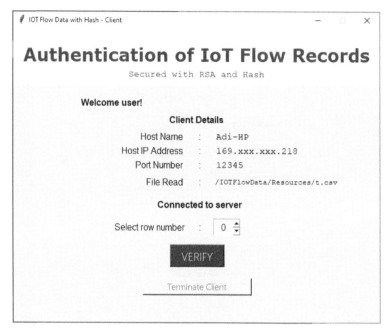

Figure 7.17 Client server establishment using RSA with hash-server side. Source: The PHP Group.

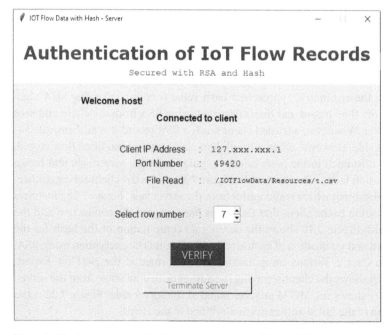

Figure 7.18 Server responding to the client. Source: The PHP Group.

Figure 7.19 Hash authenticated at the server side. Source: The PHP Group.

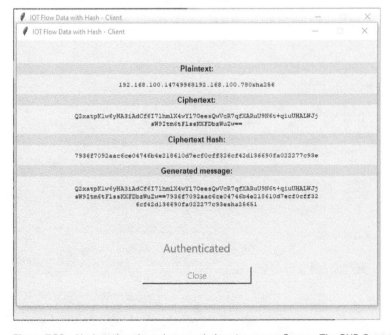

Figure 7.20 Hash authenticated responded to the server. Source: The PHP Group.

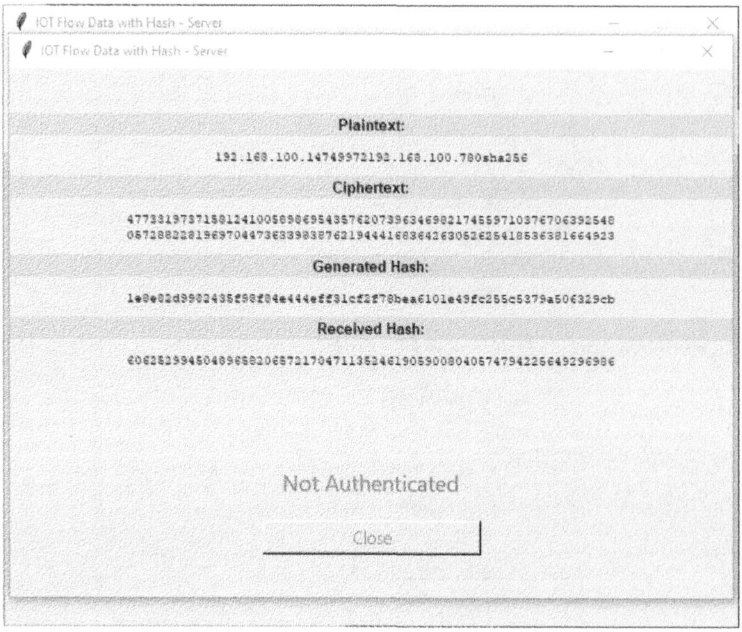

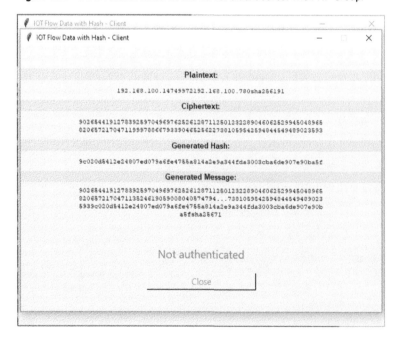

Figure 7.21 MITM attack failed at the server side. Source: The PHP Group.

Figure 7.22 MITM attack failed – client notification. Source: The PHP Group.

Table 7.1 Attributes of flow record.

Packet seq Id

Protocol

Src address

Source port

Destination address

Destination port

Seq

Standard deviation

Mean

Destination rate

Source rate

Max

Attack type

Category

Subcategory

The Cyber Range Lab of the center of UNSW Canberra Cyber[1] has released a BoT-IoT dataset which is used for our analysis. The different attacks included in the dataset are DDoS, DoS, Data exfiltration, and Keylogging attacks. Only 5% of the original dataset is extracted by MYSQL select queries and used for our analysis [60]. The different attributes in the data are shown in Table 7.1. A simulative analysis of BoT-IoT dataset is performed, and three different authentication mechanisms have been proposed and developed. Each case ensures the integrity of the flow record and identifies malicious flow records by verifying the hash digest at the NetFlow Export. The forged flow records will result in a new hash value if the attributes of flow record are modified by illegal client. The new hash value will not match the original hash value sent by the authenticated client. Thus, a robust authentication mechanism is developed to prevent the tampering of flow records by malicious devices in the IoT network. The experiments and results of the simulation of this chapter are available in GitHub [61].

7.6 Conclusion

Identifying threats in encrypted network traffic without compromising the integrity of the data is an important issue. Many IoT devices communicate with

---

1 https://www.unsw.adfa.edu.au/unsw-canberra-cyber/cybersecurity/ADFA-NB15-Datasets/bot_iot.php.

the cloud to store and process information. Telemetry of this network flow is used to analyze the traffic. Generally this flow telemetry is not encrypted. An attacker can obtain sensitive details about the IoT device, namely its type and explicit behavior of IoT traffic. Hence, IP flow data can be modified to attack the NetFlow Collector. In this chapter, we have discussed different flow-based approaches for traffic classification and how to ensure integrity of flow records between senders to collector in IoT network. Hashing and encryption methods are considered as a means for ensuring message authentication in the literature. Hence, we have proposed hashing-based and encryption-based mechanisms for flow record authentication to ensure integrity of the flow records. We have considered three different scenarios for flow-based authentication to ensure the integrity of flow records. A benchmark IoT dataset is utilized for experimental analysis to show the effectiveness of the model. Different cases are implemented to show that when an attacker tries to forge the flow records, there will be no compromise. Recently, an analysis on encrypted network traffic is conducted and is available at [62].

List of Abbreviations

AES   Advanced Encryption Standard
DES   Data Encryption Standard
DPI   Deep Packet Inspection
IDEA   International Data Encryption Algorithm
IDS   Intrusion Detection System
SVM   Support Vector Machine

Acknowledgment

This study was undertaken with the support of the Scheme for Promotion of Academic and Research Collaboration (SPARC) grant SPARC/2018-2019/P616/SL "Intelligent Anomaly Detection System for Encrypted Network Traffic."

References

1 Sivanathan, A., Gharakheili, H.H., and Sivaraman, V. (2020). Managing IoT cyber-security using programmable telemetry and machine learning. *IEEE Transactions on Network and Service Management* 17: 60–74.

2 Sivanathan, A., Gharakheili, H.H., Loi, F. et al. (2018). Classifying IoT devices in smart environments using network traffic characteristics. *IEEE Transactions on Mobile Computing* 18 (8): 1745–1759.

3 Suo, H., Wan, J., Zou, C., and Liu, J. (2012). Security in the Internet of Things: a review. *Proc. Int. Conf. Comput. Sci. Electron. Eng. (ICCSEE)* 3: 648–651.

4 Jing, Q., Vasilakos, A.V., Wan, J. et al. (2014). Security of the Internet of Things: perspectives and challenges. *Wireless Netw.* 20 (8): 2481–2501.

5 Andrea, I., Chrysostomou, C., and Hadjichristofi, G.C. (2015). Internet of Things: security vulnerabilities and challenges. *Proc. IEEE Symp. Comput. Commun. (ISCC)* (July 2015), pp. 180–187.

6 Apthorpe, N., Reisman, D., Sundaresan, S. et al. (2017). Spying on the smart home: privacy attacks and defenses on encrypted IoT traffic. arXiv preprint arXiv:1708.05044.

7 Atmani, A., Ibtissame, K., Nabil, H., and Chaoui, H. (2019). Big data for internet of things: a survey on IoT frameworks and platforms. *International Conference on Artificial Intelligence and Symbolic Computation*, pp. 59–67. Springer, Cham.

8 Guardian, T. (2016). Why the internet of things is the new magic ingredient for cyber criminals. [Online]. https://goo.gl/MuH8XS.

9 PwC (2015). Information security breaches survey. [Online]. https://www.pwc .co.uk/assets/pdf/2015-isbstechnical-report-blue-03.pdf (accessed 26 March 2016).

10 Yu, T., Sekar, V., Seshan, S., Agarwal, Y., et al. (2015). Handling a trillion (Unfixable) flaws on a Billion devices: Rethinking network security for the internet-of-things. *Proc. 14th ACM Workshop Hot Topics Netw.* (November 2015), Art.no. 5.

11 Sivanathan, A., Sherratt, D., Gharakheili, H. H., Sivaraman, V., et al. (2016). Low-cost flow-based security solutions for smart-home IoT devices. *Proc. IEEE Int. Conf. Advanced Netw. Telecommun. Syst.* (November 2016), pp. 1–6.

12 Miettinen, M., Marchal, S., Hafeez, I. et al. (2017). IoT sentinel: Automated device-type identification for security enforcement in IoT. *2017 IEEE 37th International Conference on Distributed Computing Systems (ICDCS)*, pp. 2177–2184. http://dx.doi.org/10.1109/ICDCS.2017.283.

13 Maiti, R.R., Siby, S., Sridharan, R., and Tippenhauer, N.O. (2017). Link-layer device type classification on encrypted wireless traffic with COTS radios. In: *ESORICS 2017: 22nd European Symposium on Research in Computer Security, Proceedings* (eds. S.N. Foley, D. Gollmann and E. Snekkenes), 247–264. Springer International Publishing, Oslo, Norway http://dx.doi.org/10.1007/978-3-319-66399-9_14.

14 Svoboda, J., Ghafir, I., and Prenosil, V. (2015). Network monitoring approaches: an overview. *International Journal of Advances in Computer Networks and Its Security* 5 (2): 88–93.

15 Merkle, L.D. (2008). Automated network forensics. *Proceedings of the conference on genetic and evolutionary computation (GECCO 2008)*, pp. 1929–1932.

16 Wang, K. and Stolfo, S.J. (2004). Anomalous payload-based network intrusion detection. *Proceeding of the 7th International Symposium RAID*, pp. 203–222, Sophia Antipolis, France (September 15–17 2004), in press.

17 QOSMOS (2015). Security information and event management (SIEM) use case. [Online]. http://www.qosmos.com/wpcontent/uploads/2015/08/Qos mos_SIEM_Use-Case_2015.pdf (accessed 26 March 2016).

18 Claise, B. Cisco systems NetFlow services export version 9 (2004). [Online]. https://tools.ietf.org/html/rfc3954.

19 Sflow. sflow. [Online]. www.sflow.org (accessed 7 October 2021).

20 Trammell, B. and Claise, B. (2013). Flow aggregation for the IP flow information export (IPFIX) protocol. [Online]. https://tools.ietf.org/html/rfc7015 (accessed 21 April 2015).

21 Li, B., Springer, J., Bebis, G., and Gunes, M.H. (2013). A survey of network flow applications. *Journal of Network and Computer Applications* 36 (2): 567–581.

22 Swire, P., Hemmings, J., and Kirkland, A. (2016). Online privacy and ISPs. *The Institute for Information Security & Privacy*. https://www.iisp.gatech.edu/sites/default/files/images/online_privacy_and_isps.pdf.

23 Stallings, W. (2006). *Cryptography and Network Security, 4/E*. Pearson Education India.

24 Apthorpe, N., Reisman, D., Sundaresan, S., Narayanan, A., et al. (2017). Spying on the smart home: Privacy attacks and defenses on encrypted IoT traffic. arXiv preprint arXiv:1708.05044.

25 Hofstede, R., Čeleda, P., Trammell, B. et al. (2014). Flow monitoring explained: from packet capture to data analysis with netflow and ipfix. *IEEE Communications Surveys & Tutorials* 16 (4): 2037–2064.

26 Velan, P., Čermák, M., Čeleda, P., and Drašar, M. (2015). A survey of methods for encrypted traffic classification and analysis. *International Journal of Network Management* 25 (5): 355–374.

27 Doshi, R., Apthorpe, N., and Feamster, N. (2018). Machine learning DDoS detection for consumer Internet of Things devices. *Proceedings of the IEEE Security Privacy Workshops (SPW)*, pp. 29–35.

28 Sun, G.-L., Xue, Y., Dong, Y., Wang, D., & Li, C. (2010). An novel hybrid method for effectively classifying encrypted traffic. *2010 IEEE Global Telecommunications Conference GLOBECOM 2010*. IEEE.

29 Montigny-Leboeuf, D. (2005). Annie. *Flow attributes for use in traffic characteri-zation.* Communications Research Centre Canada.

30 Anderson, B. and McGrew, D. (2016). Identifying encrypted malware traffic with contextual flow data. *Proceedings of the 2016 ACM workshop on artificial intelligence and security.*

31 Lu, G., Zhang, H., Qassrawi, M., and Yu, X. (2012). Comparison and analy-sis of flow features at the packet level for traffic classification. *International Conference on Connected Vehicles and Expo (ICCVE),* pp. 262–267. IEEE.

32 Arndt, D.J. and Zincir-Heywood, A.N. (2011). A comparison of three machine learning techniques for encrypted network traffic analysis. *IEEE Symposium on Computational Intelligence for Security and Defense Applications (CISDA).* IEEE.

33 Alshammari, R., and Zincir-Heywood, A.N. (2009). Machine learning based encrypted traffic classification: Identifying ssh and skype. *IEEE Symposium on Computational Intelligence for Security and Defense Applications.* IEEE.

34 Alshammari, R. and Zincir-Heywood, A.N. (2011). Can encrypted traffic be identified without port numbers, IP addresses and payload inspection? *Computer Networks* 55 (6): 1326–1350.

35 Yamada, A., Miyake, Y., Takemori, K., Studer, A., & Perrig, A. (2007). Intru-sion detection for encrypted web accesses. *21st International Conference on Advanced Information Networking and Applications Workshops (AINAW'07),* Volume 1. IEEE.

36 sFlow (2019). [Online]. https://sflow.org/about/index.php (accessed 1 May 2019).

37 Levi, D., Meyer, P., and Stewart, B. (2002). Simple network management proto-col (SNMP) applications. *RFC, Rep. 3431.*

38 Srinivasan, V., Stankovic, J., and Whitehouse, K. (2008). Protecting your daily in-home activity information from a wireless snooping attack. *Proc. UbiComp,* pp. 202–211.

39 Acar, A., Hossein Fereidooni, Tigist Abera, Amit Kumar Sikder, et al. (2018). Peek-a-boo: I see your smart home activities, even encrypted!. [Online]. https://arxiv.org/abs/1808.02741 (accessed August 2018).

40 Li, Y., Miao, R., Kim, C., and Yu, M. (2016). FlowRadar: a better NetFlow for data centers. *Proc. USENIX NSDI,* Santa Clara, CA, USA (March 2016), pp. 311–324.

41 Claise, B. (2004). Cisco systems NetFlow services export version 9. IETF, RFC 3954, October 2004. [Online]. https://www.rfceditor.org/info/rfc3954.

42 Zhu, Y. Kang, N., Cao, J., Greenberg, A., Lu, G., et al. (2015). Packet-level telemetry in large datacenter networks. *Proc. Conf. Special Interest Group Data Commun (SIGCOMM).* New York, NY, USA, pp. 479–491.

43 Rasley, J., Stephens, B., Dixon, C. et al. (2014). Planck: Millisecond-scale monitoring and control for commodity networks. *SIGCOMM Computer Communication Review* 44 (4): 407–418.

44 McKeown, N., Anderson, T., Balakrishnan, H. et al. (2008). OpenFlow: enabling innovation in campus networks. *SIGCOMM Computer Communication Review* 38 (2): 69–74.

45 Duda, R.O., Hart, P.E., and Stork, D.G. (2001). *Pattern Classification*, 2e. John Wiley & Sons.

46 Williams, N., Zander, S., and Armitage, G. (2006). A preliminary performance comparison of five machine learning algorithms for practical IP traffic flow classification. *ACM SIGCOMM Computer Communication Review* 36 (5): 5–16.

47 Msadek, N., Soua, R., and Engel, T. (2019). IoT device fingerprinting: machine learning based encrypted traffic analysis. In: *IEEE Wireless Communications and Networking Conference (WCNC)*, 1–8. IEEE.

48 Juniper (2015). Juniper flow monitoring. [Online]. http://www.juniper.net/us/en/local/pdf/appnotes/3500204-en.pdf (accessed 25 February 2015).

49 Healy, S.J., Quiles, S., and Von Arx, J.A. (2007). Cryptographic authentication for telemetry with an implantable medical device. US Patent 7, 228, 182. 5 June 2007.

50 Claise, B., Trammell, B., and Aitken, P. (2013). Specification of the IP flow information export (IPFIX) protocol for the exchange of flow information. *RFC 7011*.

51 Gu, G., Porras, P.A., Yegneswaran, V. et al. (2007). Bothunter: detecting malware infection through IDS-driven dialog correlation. *USENIX Security Symposium*, Volume 7, pp. 1–16.

52 Gu, G., Perdisci, R., Zhang, J., and Lee, W. (2008). BotMiner: clustering analysis of network traffic for protocol-and structure-independent botnet detection. *USENIX Security Symposium*, Volume 5, pp. 139–154.

53 Moore, A.W. and Zuev, D. (2005). Internet traffic classification using Bayesian analysis techniques. *ACM SIGMETRICS Performance Evaluation Review*, Volume 33, pp. 50–60. ACM.

54 Nagaraja, S., Mittal, P., Hong, C.-Y. et al. (2010). BotGrep: finding P2P bots with structured graph analysis. *USENIX Security Symposium*, pp. 95–110.

55 Claise, B., Sadasivan, G., Valluri, V., & Djernaes, M. (2004). RFC 3954: cisco systems NetFlow services export version 9. *IETF*. http://www.ietf.org/rfc/rfc3954.txt.

56 Nguyen, T.T. and Armitage, G.J. (2008). A survey of techniques for internet traffic classification using machine learning. *IEEE Communications Surveys and Tutorials* 10 (1–4): 56–76.

57 SolarWinds. (2010). NetFlow basics and deployment strategies. [Online]. Available: https://bit.ly/2K4EZ6Q.

58 Gu, G., Zhang, J., and Lee, W. (2008). BotSniffer: Detecting Botnet Command and Control Channels in Network Traffic. Proceedings of the 15th Annual Network and Distributed System Security Symposium. https://corescholar.libraries.wright.edu/cse/7.

59 Meidan, Y., Bohadana, M., Shabtai, A. et al. (2017). Detection of unauthorized IoT devices using machine learning techniques. arXiv preprint arXiv:1709.04647.

60 Koroniotis, N., Moustafa, N, Sitnikova, E, and Turnbull, B. (2018). Towards the development of realistic botnet dataset in the internet of things for network forensic analytics: Bot-IoT dataset. https://arxiv.org/abs/1811.00701.

61 Cherukuri, A.K., Thaseen, I.S., Li, G. et al. (2021). IOTFlowData. San Francisco (CA): GitHub. https://github.com/adityaDel222/IOTFlowData (accessed 2021).

62 Ikram, S., Cherukuri, A., Poorva, B. et al. (2021). Anomaly detection using XGBoost ensemble of deep neural network models. *Cybernetics and Information Technologies* 21 (3): 175–188. https://doi.org/10.2478/cait-2020-0037.

8

Securing Contemporary eHealth Architectures: Techniques and Methods

Naeem F. Syed[1], Zubair Baig[2], and Adnan Anwar[2]

[1] School of Science, Edith Cowan University, Perth, Australia
[2] School of Information Technology, Deakin University, Geelong, Australia

8.1 Introduction

The emergence of the Internet of Things (IoT) paradigm is transforming the way healthcare services are provided to patients especially in eHealth which is reliant upon digital resources to deliver efficient and quality healthcare. The ubiquitous nature of IoT devices allows continuous monitoring and real-time collection of patient health data, facilitating a timely medical intervention from healthcare professionals. This enhances efficiency and delivery time along with providing a personalized and cost-effective healthcare services to patients to improve quality of life as well as longevity. The wide range of IoT sensor devices comprising implantable, wearable, wired, and wireless devices provide a rich source of medical data that enables healthcare professionals to perform early detection of diseases and to provide a data-driven and intelligent treatment plan to facilitate successful rendering of treatment. The inclusion of IoT devices makes the eHealth architectures complex and sophisticated; however, the benefits of these technological advances can be reaped to render effective and efficient patient services. The numerous use cases and benefits of IoT technology in eHealth are accompanied with an ever-increasing landscape of cyber-security threats posed. The primary reason being the increased connectivity between the eHealth devices and the lack of proper mechanisms for ensuring security of IoT devices of the eHealth system. In order to identify the various threats to the eHealth system and possible countermeasures, we first define the eHealth system and the role of IoT and its numerous applications in healthcare (Section 8.2). A threat landscape of IoT-based medical systems which describe the various assets of the system is presented, followed by an explanation of vulnerabilities and threats for various

Security and Privacy in the Internet of Things: Architectures, Techniques, and Applications,
First Edition. Edited by Ali Ismail Awad and Jemal Abawajy.
© 2022 The Institute of Electrical and Electronics Engineers, Inc. Published 2022 by John Wiley & Sons, Inc.

digital assets present within individual layers of an eHealth system (Section 8.3). This is followed by a detailed analysis of countermeasures to protect eHealth systems from cyber-attacks (Section 8.4). The chapter is then concluded through a summary of findings presented thereof (Section 8.5).

8.2 eHealth

The term "eHealth" includes a broad spectrum of healthcare services that depend on Information and Communication Technologies (ICT) to deliver personalized care to patients. The main aim of an eHealth system is to bridge the gap between various beneficiaries and stakeholders of healthcare systems and to provide a patient-centered personalized service by effectively adopting and deploying ICT. The various services offered under a patient-centered eHealth umbrella include electronic patient medical records, remote (also referred to tele) health care services, virtual healthcare teams, and medical research and health informatics. In patient-centered healthcare system, patients own their health records and can further authorize various healthcare stakeholders to provide targeted health services [1]. The chronologically ordered health records of patients stored and processed electronically are referred to as Electronic Health Record (EHR), which is one of the key enablers of eHealth. Figure 8.1 shows the various EHR-based services comprised in a patient-centered healthcare system. Healthcare providers can adopt the EHR to develop a patient-centered care, to perform accurate diagnosis, and to render personalized medicine based on the patient's medical history and current medical conditions [1]. Another aspect of eHealth is that the geolocation of patient as well as the healthcare professional involved in the treatment does not hinder the delivery of personalized care to the patient. The accessibility of EHR from any location, remote or in proximity, enables patient-centered care to be provided within such a system.

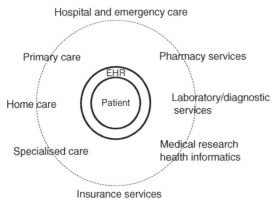

Figure 8.1 Various services provided in a patient-centered eHealth ecosystem.

Hospital and emergency care
Primary care
Pharmacy services
EHR
Home care
Patient
Laboratory/diagnostic services
Specialised care
Medical research health informatics
Insurance services

The introduction of ICT and the use of EHR in healthcare systems brings along numerous advantages such as [1, 2] automated collection of data and increased access to patient's medical records, convenience and support on demand as well as coordinated care by various healthcare providers, flexibility and modifiability of healthcare programs, to enable global access of medical data, and to help reduce interoperability issues, ratio of erroneous treatments and delays in rendering treatment, remote delivery of healthcare, low cost, and effective and tailored healthcare. The rich healthcare information generated in a patient-centered health system brings along numerous other opportunities for subsidiary health services such as medical research, pharmacy services, laboratory and diagnostic services, and insurance services as highlighted in Figure 8.1.

The health data generated by various monitoring devices in close proximity to patients will be crucial for the realization of the abovementioned advantages and services [3]. This has been made possible due to the various technological advancements and innovations in sensor network technologies and growing hardware capabilities which have led to rapid development of IoT. The ability of IoT devices to communicate anywhere and anytime has opened up many opportunities in the eHealth sector. In addition, the ability to monitor a patient's health using implantable and wearable IoT devices has led to new dimensions appertaining to patient data, that can be included in an electronic medical record (EMR), to enable quick and personalized treatments [4]. In the following section, the role and advantages of IoT in the eHealth domain are elaborated.

8.2.1 Why IoT Is Important in eHealth?

A connected medical and health environment requires secure and faster communication of information. Besides this, information also needs to be processed efficiently and reliably. With the recent advances in Cloud and edge technologies, it is possible to store and rapidly analyze millions of patient data records in the Cloud environment. IoT has also opened up enormous possibilities for sensing, storing, aggregating, and analyzing data within Cloud or edge nodes. Medical data can be transmitted in real time using IoT connected end-devices and processed either at the edge or within Cloud nodes based on the stakeholder requirements. The connected IoT-based eHealth not only ensures efficient and faster medical data processing but also encourages an evidence-based system to be adopted for rendering medical treatment. Figure 8.2 shows a typical IoT-based eHealth architecture which consists of a sensing layer, edge or Fog layer, and cloud computing layer. The various benefits of using IoT in eHealth as categorized by Firouzi et al. [4] are presented in Table 8.1.

In recent years, the importance of IoT in eHealth has been highlighted through IoT-empowered technological and technical innovations in the health domain. Some of these include advancement in smart wearable device technology, progress

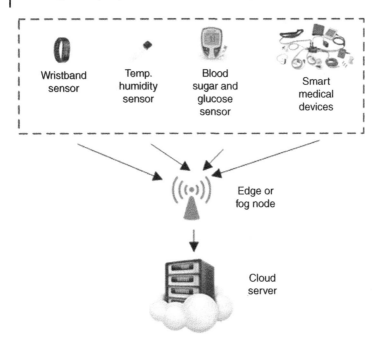

Figure 8.2 A Typical eHealth Architecture. Source: Contour Next EZ Meter, Kodea.

on intelligent and smart health data collection in real time, patient's health condition monitoring using improved medical data aggregation and fusion within IoT systems, development of smart biosensors, utilization of Cloud facilities for storage and computing of millions of patient data, precision robotic surgery using smart sensors and actuators, and remote and distant medical treatment using IoT-connected advanced communication.

8.2.2 Fog or Edge Computing for eHealth

Fog or edge computing extends the capability of Cloud computing by bringing the computational and analytical capabilities near to the edge of the network. In an eHealth setup, sensors, medical devices, and gateways are connected to the Fog nodes, which are basically responsible for medical data aggregation, processing, and transmission to the Cloud node. As the edge or Fog node is inserted in between the Cloud and the patient's data source, as it is equipped with lots of advantages, such as [4]:

- Real-time analytics and decision-making based on patient data.
- Transients data processing of any medical devices for shorter period is possible based on Fog data analytics.

Table 8.1 Various benefits with the use of IoT in eHealth.

Advantages	Description
All-encompassing	Use of IoT in eHealth has potential benefits for various end users, from patients to healthcare providers. Patients have more control over their health and fitness status, and healthcare providers get more streamlined operations, automation, and cost-saving benefits
Resiliency	An adaptive eHealth framework will quickly adapt to changes and be resilient to errors
Seamless fusion with different technologies	Various technologies can combine seamlessly in a complex eHealth system to improve rendering of healthcare services
Big data processing and analytics	Range of IoT sensor devices provide a rich set of healthcare datasets thereby enabling deeper insights into diseases and their treatments
Personalized forecasting	The valuable data acquired about the patient's health parameters can allow forecasting of future health problems and can enable early intervention planning
Lifetime monitoring	A comprehensive health report of patients can be generated providing insights into their past, present, and future health conditions
Ease of use	IoT-based eHealth solutions will provide convenient-To-use wearable devices and adaptive mobile applications
Cost reduction	Streamlining of processes, avoiding duplication, automation of services will help reduce operational costs
Physician oversight	Allowing automated health information collection reduces the time spent by doctors in treating individual patients thereby allowing them to treat more patients
Availability and accessibility	Anytime and anywhere IoT access allows for a round the clock health services to be facilitated
Efficient healthcare management	Detailed IoT populated EHR records provide an efficient healthcare management for both patients and providers
International impact	A connected world of patients and doctors (domestic and international) provide patients a specialized care which would otherwise be delayed or inaccessible

- Edge node can process data cleaning, data filtering, compression, data wrangling, etc. Moreover, it is possible to enhance the edge analytics by using signal processing, event handling, etc.
- Advanced data security can be obtained by introducing authentication, encryption, and access control at the Fog layer of the eHealth architecture.

8.2.3 Cloud Computing for eHealth

In an eHealth setup, the concept of *big data* refers to the huge amount of medical patient's data which are generated from a wide variety of sources, including medical sensors, end-user apps, heal-service providers, etc., in order to enhance the efficiency and reliability of the medical-surgical and diagnosis process. These large numbers of generated data are not only huge in volume but also satisfy other big data criteria like variety and velocity. As a result, storing these huge amounts of data, maintaining time-requirement, accessing shared resources and analysis, and processing of these data is a very challenging task. Cloud computing is a promising solution and executes most of these operations [5].

In [6], authors highlight the initiative by the National Institutes of Health (NIH) which announced the "All of Us" program. As a part of this initiative, the EHR of more than one million patients will be collected. Similar to EHR, EMR also includes data related to the patient's medical and clinical conditions. Other forms of data such as pharmacy prescription and insurance records along with the EHR and EMR combine a complete big data environment for the medical system. As the Cloud is a promising solution, most of the data storage and processing are performed at the Cloud. The authors in [6] present a Could-based architecture where the process of data generation, transmission, and processing in an eHealth system under the big data environment are elaborated.

8.2.4 Applications of IoT in eHealth

eHealth has been benefited significantly with the adoption of IoT-enabled technologies. The range of applications for IoT in healthcare includes continuous patient health monitoring, in-hospital care, and long-term posttreatment care. The IoT devices can be deployed to measure and report the people's vital health parameters, equipment used in medical industry, and biometric information required for medical diagnosis [7]. A handful of IoT application in eHealth is as follows:

8.2.4.1 Sleep Monitoring System

An IoT-enabled smart sleep monitoring system is presented in [8]. The scheme adopts the capability of Fog-computing at the preprocessing stage, whereas batch data processing is accomplished at the Cloud level. The authors emphasize the importance of air quality monitoring for the treatment of Obtrusive Sleep Apnoea (OSA) [8]. The authors also suggest an IoT-based predictive model for this purpose. In [9], the author's propose an agent-based IoT simulation testbed for posture detection during the sleeping. This model helps diagnose sleeping misbehavior patterns.

8.2.4.2   Real Time and Advanced Health Diagnoses

Health diagnoses can be significantly improved by incorporating patient data that may emerge from multiple sources. Realizing the fact of the data-driven decision-making toward an improved health diagnosis system, authors in [4] propose an innovative model that can capture the medical data produced by a large number of IoT-enabled medical sources. In [4], authors also highlighted that a model-based diagnosis approach for a medical system may not be suitable because of the changing nature of the medical data. Hence, an adaptive and data-driven model may be a better solution that is realizable through the adoption of IoT devices.

8.2.4.3   Emotion Detection

In [10], the authors propose an emotion detection model that uses IoT-enabled speech and image signals. Signal processing and machine learning capabilities are adopted for analyzing the data in order to detect emotional changes.

8.2.4.4   Nutrition Monitoring System

Another IoT-based application is a nutrition monitoring system. In [11], authors utilize Bayesian models for meal-prediction to ensure meeting the minimum nutrition criteria. Deep learning-based another model is also proposed to balance the nutrition requirements. The complete process is an IoT-driven initiative.

8.2.4.5   Detection of Dyslexia

Distributed ledger technologies such as blockchains can be adopted for the treatment of dyslexia, which is a cognitive disability that can be detected in an early stage with the aid of IoT by mining data from smart devices such as smart phones, tabs. Some other applications of IoT in eHealth include as below [4]:

- Personalized forecasting of patient condition
- Use of wearable devices
- Lifetime monitoring of patient health
- Remote medical access
- Patient appointment scheduling
- Precision surgery
- Smart continuous glucose monitoring
- Smart coagulation testing
- Enhanced drug management
- Improved patient data and record management

8.2.5   eHealth Security

Numerous benefits can be realized for an eHealth system, based on adoption of automation, IoT, and artificial intelligence techniques. The most prominent benefit is the delivery of a localized, tailored, and personalized healthcare service to patients by continuously monitoring patient health conditions. However, the introduction of ICT into the healthcare industry also opens the door for major cyber security threats, mainly associated with the sensitive nature of data being exchanged between patients and care providers. According to a report [12] that was based on a survey conducted with respondents from various organizations around the world, healthcare was the second highest targeted sector for cyberattacks after the government sector. One of the reasons for healthcare being one of favorite target for cyber criminals is that the medical records fetch high monetary value in the Darknet (a cyber underworld), when compared to other sensitive data [12, 13].

Furthermore, the threats to healthcare systems have been increasing with the introduction of IoT in healthcare applications. Apart from the monetary benefits of patients' medical records, the proximity of most eHealth-based IoT devices to patients due to its application in health monitoring, drug dispensing, treatment of chronic diseases, and fitness program management lures adversaries to misuse the data so as to cause harm to the patients. In order to identify various threats to the eHealth system, we first define the threat landscape, the various assets that might be targeted, and the attack agents. Second, the vulnerabilities of various assets and their associated threats are discussed. Finally, some of the prominent real-world cyberattacks to eHealth are also discussed to highlight the security problems faced by the healthcare sector.

8.2.5.1 Implications of eHealth Security for Smart Cities

With the IoT paradigm gaining rapid advances into daily routines of citizens of a nation, a smart city is thus empowered; a plethora of applications and technologies now govern a smart city's technology-driven activities ranging from smart transportation to smart buildings and the likes [14]. IoT devices embedded within hospital systems range from patient management all the way to telemedicine, wherein the patients are operated upon remotely through IoT-enabled and virtual reality-driven platforms. The security of such sophisticated eHealth systems in a smart city environment is lacking, and if it does exist, has questionable efficacy.

With more than one billion people over the age of 60 living in 2020, the implications of a security compromise on an eHealth system are both significant as well as underestimated [15]. Wearable devices are increasingly being deployed on patients to aid in monitoring of health and vital signs including blood pressure, oxygen saturation levels, etc. These IoT devices not only collect the patient's vital parameters but also transmit the readings through wireless channels to Fog or centralized Cloud storage facilities for data analytics, storage, and further transmission purposes. Patient data are mostly sensitive even if patients have given their consent to privacy-preserving and sharing of their obfuscated data. However, security is not given due importance during the design process of such eHealth systems of smart cities. This necessitates a comprehensive analysis of threats to the eHealth system and identifies potential vulnerabilities and threats to such systems. Such an analysis will pave the way for building appropriate countermeasures for a sensitive and critical healthcare system. In the following sections, we identify the various components of an eHealth system and the critical assets that need to be secured. We then elaborate the various vulnerabilities and security threats of these assets which are necessary to identify the appropriate countermeasures.

8.3 eHealth Threat Landscape

In a vast healthcare ecosystem, IoT can be leveraged for a wide range of applications. A typical IoT-based eHealth system comprises IoT devices, gateways, and Hospital Information Network (HIN). IoT devices are deployed for monitoring the patients, medical equipment, and medical resources. The data collected by IoT devices is the main source of information as fed into an EHR [16]. IoT devices run various protocols to enable exchange of data sensed by them based on their communication capabilities. Some IoT devices would require an intermediate gateway device to transfer the sensed patient's data to centralized application servers. The gateways can include Cloud-based or Fog-computing nodes connecting the medical IoT devices to the HIN. The HIN will connect care givers to the patient as well as the various departments and partners who contribute in providing care to the patients.

The connected IoT devices at the patient's end form a Body Area Network (BAN), which can be further classified into implantable, wearable, unconnected, or connected devices based on a medical application use-case [17]. These devices can be configured to send real-time continuous monitoring data of critical patients to centralized processing nodes (Cloud or Fog), whereas a discrete or one-time data transmission can be adopted for noncritical patients [17]. In terms of the connectivity, devices in the BAN can be categorized into wired or wireless devices. Certain devices such as implanted or wearable devices might require gateways to transmit data to the Cloud or Fog nodes. Due to limited processing capabilities, these medical devices require intermediate devices (gateways) that can enable inter-BAN communications as well as communication with the Internet [18]. These intermediate devices support various IoT protocols to establish interoperability and also provide mechanisms to securely communicate on the Internet.

8.3.1 eHealth Threat Model

In order to understand the potential threats to an eHealth system, its various assets and attack agents need to be identified. The assets are the primary entities that are targeted to breach the system and thereby needs attention to secure them. Various attack agents with varied attack motivations form the adversary base of the system.

8.3.1.1 eHealth Assets

The primary assets that the adversaries target in a healthcare system are either related to the patient or the hospital [19]. The primary patient assets include patient health and health records. Similarly primary hospital assets include reputation, Intellectual Property (IP), finances, and physician reputation. In order to breach these assets, various physical assets in an eHealth system can be threatened by a cyber-attack. The main physical assets of an IoT-based eHealth solution that might become a target of a cyber-attack are [20]:

- Medical IoT devices
- Smartphones and other intermediary devices
- Smart gateways, Fog nodes
- Cloud nodes
- Hospital Information Network
- Partner Network

According to [17], data in the eHealth ecosystem is the primary motivation for adversaries; however, for cyber defenders patient health is of the highest priority [19]. The potential entry points to eHealth network are provided by attack surfaces such as the medical IoT devices, patient-end IT devices, HIN and patient-end and hospital communication networks, gateways, middle-ware platforms, Fog and Cloud infrastructure, and partner networks as highlighted in Figure 8.3.

8.3.1.2 eHealth Attack Agents

The potential agents that can target eHealth systems can be categorized into internal and external agents [21]. Internal threat sources can be due to internal users such as

- *Patient (end user).* Both in-patient and remote patients can become an internal source of attacks and their primary motivation could be to falsify medical records.
- *Medical staff (data users).* These users can engage in tampering data and health records to cover up for fraudulent health practices.
- *Technical staff (manufactures, disgruntled employees).* Manufactures could deliberately install backdoor tools on their manufactured IoT devices, so as to access device data as well as for remote control during troubleshooting. Other internal threat agents could be disgruntled hospital employees having access to devices and data, getting involved in damaging the reputation of their employers.

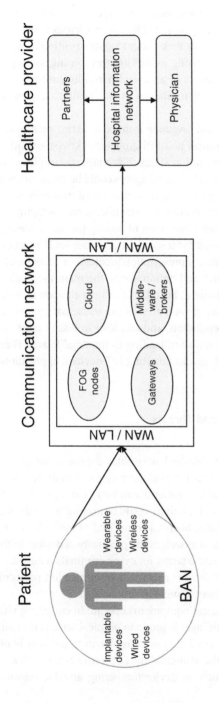

Patient

BAN

Implantable devices
Wired devices
Wearable devices
Wireless devices

Communication network

WAN / LAN

FOG nodes
Cloud
Middle-ware / brokers
Gateways

WAN / LAN

Healthcare provider

Partners
Hospital information network
Physician

Potential attacks
• Jamming
• Firmware reverse engineering
• Side-channel attacks
• Device tampering
• Resource exhaustion
• Selective forwarding
• SE attacks

Network attacks directly targeting IoT devices
• Malware
• Authentication attacks
• DoS
• Replay attacks
Network attacks targeting protocols, middle-wares, gateway
• Network layer DoS
• Application layer DoS
• Man in middle attacks
• Gateway attacks

DoS
Data breach
SQL injection
Malware
SE attacks

Figure 8.3 A typical eHealth architecture showing some of the prominent assets in various layers.

Among the internal and external attack agents, internal agents pose a higher risk to eHealth systems due to their access rights [22, 23]. The malicious insider can potentially alter medical records, delete, or leak patient's data resulting in wrong diagnosis and treatment of the patient causing serious life-threatening issues [23]. Insider attacks can occur at various layers of an eHealth architecture targeting physical IoT sensor devices in the BAN to attacks in the healthcare provider network.

In contrast, external attack threats can originate from adversaries such as [19] individual malicious user or small groups, political groups/hacktivist, organized criminals, terrorist organizations, and nation states. The main motivation of an individual or a small group and organized criminal agents could be monetary benefits from accessing sensitive patient's data, such as credit card numbers, social security numbers, and insurance identifiers [19, 24]. Healthcare data is highly valued in the blackmarket and could fetch large sums of money for stolen medical records [13]. The stolen data can be used for fraudulent activities such as accessing medical assistance, insurance claims, or prescription medication. However, for political groups, terrorist organizations, and nation states, the main motivations could be to cause harm to either the patient, hospitals, or the healthcare industry in general [13]. The adversarial impact of such attacks can cause patient deaths, tarnishing a medical practitioner's reputation, and can lead to loss of trust and reputation damage in the healthcare industry, leading to financial losses. These attack surfaces and agents can attract attacks targeting IoT devices, applications, and the communication channel.

8.3.2 eHealth IoT Vulnerabilities and Threats

8.3.2.1 Attacks in BAN

The various sensor devices (implantable and wearable) deployed for patient monitoring and delivering healthcare to patients along with gateway devices form a BAN [22]. Both remote and hospitalized patients can be in close proximity to many IoT devices. According to a report [25], there could be around 10–15 devices per bed in the United States with a large hospital comprised of more than 5000 beds. Such a large deployment of devices as well as the proximity of devices to the patients, makes these devices an attractive target for cyber criminals who intend to cause damage. Specifically, the IoT devices and the communication protocols used in a BAN can be the primary targets of such cyberattacks [26].

As numerous devices (IoT and medical equipment) operate in collecting vital health information of the patients, this layer is prone to physical attacks that aim to either damage the device, alter or steal data or cause harm to patients being monitored. Adversaries can exploit the vulnerabilities of hardware, of software interfaces, and perpetrate attacks such as device-tampering attacks, resource exhaustion attacks, side-channel attacks, and firmware reverse engineering attacks, which can lead to device compromise, unauthorized control, and can be further leveraged to target the other layers of an eHealth architecture.

In addition, the communication protocols used in BAN such as RFID, NFC, Bluetooth, Z-Wave, which provide short distance communications can also be prone to attacks such as man-in-the middle attacks, side channel attacks, jamming, and selective forwarding attacks [27]. For example, communication with passive radio-frequency identification (RFID) devices can be targeted with confidentiality attacks or cause the device to fail by introducing interference. near field communication (NFC) can be targeted to cause Denial of Service (DoS) or data manipulation by signal corruption. Z-wave-based devices are prone to identify spoofing, BlackHole, and impersonation attacks [27]. Bluetooth devices can compromise of firmware modification, software injection, device ID spoofing, and man in the middle attack (MiTM) attacks by sniffing Bluetooth communication.

The patients themselves can become victims of cyber-attacks of crafted social engineering attacks resulting in sensitive healthcare data loss. Such attacks aim to lure the patients by baiting or pretexting to reveal their sensitive information such as authentication passwords or disguise as support technician to gain access to implantable devices resulting in unauthorized access to these devices.

8.3.2.2 Attacks in Communication Layer

The communication layer of an eHealth architecture will comprise gateways, middleware, Fog, and Cloud nodes. This layer can be susceptible to various network-based attacks such as Network Layer and Application Layer DoS attacks, that may attempt to exhaust computing and communications resources of the ICT systems. Man-in-the-middle attacks can pose security threats to the integrity of the data exchanged between healthcare IoT devices and the hospital network [28]. In addition, the Fog and Cloud infrastructure can become targets of cyberattacks such as advanced persistent threats, access control attacks, account hijacking, DoS, data, and application attacks [29]. Many of communication layer protocols such as Message Queuing Telemetry Transport (MQTT), Hyper-Text Transfer Protocol (HTTP), and Constrained Application Protocol (CoAP) are also vulnerable to Application layer attacks such as MiTM attacks, data interception and manipulation attacks, and DoS attacks [27].

8.3.2.3 Attacks in Healthcare Provider Layer

The hospital network will collect and process data from various patients and other sources to provide efficient healthcare to end users. As many stakeholders are involved in dealing with patients' medical records, there are numerous threats to this layer in the eHealth system [19, 28]. The various eHealth applications deployed in the healthcare provider layer will be vulnerable to application attacks to cause disruption of services or to access sensitive healthcare information. The various software interfaces that are used to exchange patient data among multiple hospital software systems can be prone to various cyberattacks that can result in the failure of the applications. Attacks such as code injection, structured query language (SQL) injection, script injection, and data disclosure especially on the EHR systems can be launched. The hospital network can also be prone to targeted attacks in a deliberate attempt to damage the hospitals reputation and to steal sensitive patient data for financial gains by holding the hospital hostage using ransomware attacks. Attacks can also target partner network or physicians to gain access to hospital network.

The resultant threats and possible attacks to IoT-based eHealth can be [20, 30] categorized based on the layers of the eHealth architecture and are presented in Table 8.2.

The main motivations of such attacks are to target the medical data or the system used to facilitate the eHealth objectives. The main security goals that adversaries try to target are data protection and system protection which can be subdivided into user and device protection [31]. According to Yaacoub et al. [31], each of these security goals of the medical IoT systems can become targets of cyberattacks causing various cyber security issues such as sensitive patient data theft, use of medical IoT devices to launch other cybercrimes, or attack to organizations that are involved in the medical IoT ecosystem. The attacks that target the data protection are data confidentially attacks, data integrity attacks, message authentication attacks, and data availability attacks. The attacks that target the user protection are privacy attacks and user authentication attacks. Lastly, the device protection faces integrity, authentication, and availability attacks. These attacks highlight

Table 8.2 Cyber attack classification in eHealth system

Architecture layer	Attack types
Patient (BAN)	Jamming attacks, Firmware reverse engineering, Side-channel attacks, Device tampering, Resource exhaustion, Selective forwarding, Social engineering attacks, BAN communication protocol attacks, Devices spoofing attacks, Malware, Viruses, Worms, Trojans, Spyware, Routing attacks
Communication network	Network layer DoS, Application layer DoS, Man-in-the-middle attacks, Data interception and manipulation, Gateway attacks, Fog node attacks, Smart gateway attacks, Cloud attacks
Healthcare provider	DoS attacks, EHR tampering, EHR theft, Data disclosure, SQL injection, Malware, Ransomware, Social engineering attacks

that various target security objectives face different challenges in protecting the assets in a medical IoT system. The outcomes of such cyberattacks could be catastrophic and life-threatening as patients are directly part of the system, and many of the devices are either implanted or worn by them [32]. This would also result in a loss of trust in such applications and a loss of reputation and a significant monetary loss for organizations participating in healthcare industry.

8.3.3 Real-world Attacks

Possible attacks on medical devices and its adverse consequences have been demonstrated by targeting devices such as implantable brain devices [33], cardiac defibrillators [34], and insulin pumps [35]. Such attacks can allow the adversary to control the brain function, control heart rhythms, and potentially deliver large doses of insulin leading to adverse health conditions in patients. In addition to threats to implantable and wearable devices, cyber threats can also target other medical devices as vulnerabilities were found in X-ray equipment, picture archive and communication systems (PACS), and blood gas analyzers as highlighted in [36].

Furthermore, the cyber security breaches in hospitals have increased considerably in the last decade to target patient records to cripple healthcare services [13, 24, 37]. One of the major attack was WannaCry ransomware which crippled 50 hospitals in the United Kingdom (UK) and many healthcare institutions in the United States which resulted in cancellation or delays in medical appointments and surgeries [13].

These real-world attacks highlight the direct impact they can have on the healthcare system especially on patient's health. Even though IoT-based eHealth brings tremendous benefits directly impacting humans, these systems need to be protected from cyber-attacks. Countermeasures must be in-built within all the layers of eHealth architecture to safeguard the privacy and security of patient data. The following sections discuss the various countermeasures that had been proposed in the literature to protect eHealth system.

8.4 Countermeasures

Deployment of IoT-enabled smart medical devices has significantly improved the quality and service provided by the eHealth sector. However, due to weak security measures deployed in IoT devices, smart healthcare has been an attractive target for cyber attackers. Dedicated research work is being undertaken to improve the security monitoring and analysis capabilities of IoT-enabled applications and devices. New architectures are being developed for better security and reliability.

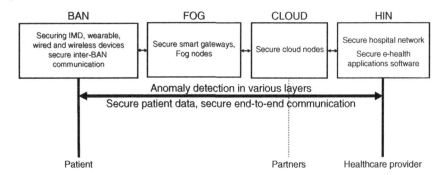

Figure 8.4 Security objectives in various layers of the eHealth architecture.

The countermeasures for a complex eHealth system require implementation of security measures at various layers of the architecture. The primary objective of countermeasures in eHealth focus on securing the medical data and preserving the privacy of patient's health information [22]. In addition, the devices that are in close proximity to the patient also need to be protected in an IoT-based eHealth system. Figure 8.4 shows the security objectives of various layers of eHealth architecture.

At the BAN layer where the IoT devices exist and are in close proximity to patients, requirements exist for security measures such as physical security of devices to prevent node capture, device tampering, and securing inter-BAN communication. Hardware security features such as Physical Unclonable Function (PUF), lightweight communication authentication protocols, and distance bound protocols for implantable devices need to be implemented at the BAN layer [32]. At the Fog layer, devices such as smart gateways and other intermediate devices that operate in the Fog need to be secured. Similarly, in the Cloud layer, nodes need to be protected from various Cloud-based attacks. Finally, in the hospital network, communication network must be secured from breaches and the application attacks need to prevented. At all the layers, attack detection and prevention mechanisms must be employed to prevent security breaches. Similarly, the sensitive patient data traversing the various layers of eHealth architecture need to be secured from adversaries. The security measures applied in the eHealth system must secure devices deployed in eHealth to prevent unauthorized device control, secure all communications occurring in the eHealth system to protect the medical data traversing through it, secure eHealth software applications from cyberattacks to prevent malicious system and data access, monitor, and detect anomalous behavior in all layers of eHealth to prevent anomalous activities and security breaches.

Based on the above, some of the countermeasures that can be applied in eHealth systems are the following:

- Patient data protection as patient data is a most critical asset in the eHealth system.
- Device and communication security measures are critical as these devices monitor vital health parameters of the patients which need to be securely transferred to the cloud servers for further processing.
- An adaptive security framework that can take feedback from all the components of eHealth system and respond to new threats as new heterogeneous devices get added to the system.

These countermeasures are further elaborated in the following sections.

8.4.1 Patient Data Protection

As the patient data exchanged in eHealth systems are confidential and sensitive, they need to be secured from cyber-attacks. The most prominent data sources in an eHealth system are the sensors generating real-time monitoring data of the patients [38]. In addition, the use of a large number of sensor devices and components in an eHealth system requires the security solution to be scalable and interoperable.

The recent advancements in crypto currencies have resulted in the development of a secure, decentralized, and peer-to-peer transaction management technology referred to as blockchain. The main advantage of blockchain technology is that it allows nontrusting members to interact with each other without the need of an intermediate trusting authority [39]. Due to the various advantages in securing the transaction data, the blockchain technology is being extended to various applications as well as to protect patient data in an eHealth system [38]. In addition to the blockchain technology, the data exchange between untrusted components in an eHealth system will be facilitated using smart contracts, which allow patients, doctors, hospitals, and insurance providers to verify communications.

Even though blockchain enables data integrity of EHRs, it does not have fine grained control on data access control [40]. Since there are multiple stakeholders who can have access to EHRs in the eHealth ecosystem, data access control is an essential component in preserving privacy of patient data. In [40], the authors have proposed a computationally efficient blockchain and elliptic curve-based, block-level access control scheme for EHR data access control and authorization in eHealth.

Authors in [41] propose a software architecture for secure and reliable IoT-enabled smart healthcare. The framework utilizes machine-to-machine (M2M) messaging with the fusion of data and decisions. Additionally, Tor and Blockchain have been used for ensuring security. Tor provides anonymity by enabling multiple random paths between communication endpoints; however, this can result in time delay during message exchange. Hence, the authors suggest using the Tor at the Fog nodes so that the time requirement is maintained. On the other hand, utilization of decentralization and distributed ledger technology like blockchain ensures the security of medical patient records. Due to the adoption of blockchain technologies, cyber attackers find it difficult to compromise the information during the data transfer in the proposed architecture. Moreover, blockchain-based decentralized architecture secures the whole system against a single point of failure during any cyber-attack.

8.4.2 Device and Communication Security Measures

Device protection measures require securing IoT devices that are implantable, wearable, or wireless devices deployed for monitoring patients in the BAN. Maintaining confidentiality and integrity are essential in a BAN due to close proximity of devices to the patients and the collection of highly sensitive data [42]. eHealth device security can be categorized into physical security of devices, authentication, and authorization of devices monitoring sensitive patient data, and securing M2M and M2C communication. One of the security measures employed in BAN is to use authentication mechanisms to prevent unauthorized IoT devices from communicating in BAN. The common authentication schemes used in IoT-based medical devices are [42]:

- *Physiological parameters.* This method uses ECG, iris, or fingerprint from the patient. In addition, to authenticate various devices to the Fog or Cloud nodes, unique device properties which exist due to manufacturing variations can be tapped to prevent spoofing and cloning attacks [43].
- *Channel-based authentication.* Channel-based authentication provides a lower-layer authentication for medical IoT devices by comparing the received signal strength as various transmitters can be differentiated based on the channel estimation mechanism. An example of light-weight authentication protocol for M2M communication is presented in [44] which uses a channel state information (CSI) obtained from the received signal to encrypt the communication session.
- *Proximity-based authentication.* Proximity-based authentication which uses devices proximity information to either grant or deny access to the device. Such authentication schemes are especially important for implantable devices which should only allow access to those users that the patient can physically see. Rasmussen et al. [45] presented a proximity-based authentication solution

which adopts the ultrasonic distance bounding to determine the distance between the reader and implantable device to allow access to the device.

In addition to the authentication schemes, devices connecting to the network must also be forced to certain security requirements to prevent unauthorized access. According to IEEE 802.15.6 standard for BAN communications [46], three levels of security are defined:

1. *Level 0.* Unsecured communication is the lowest level of security with no technique to preserve confidentiality, integrity, and privacy of the data exchange.
2. *Level 1.* Authentication only mechanism provides medium security by introducing authenticated data exchange mechanism; however, it still lacks methods to preserve confidentiality and privacy.
3. *Level 2.* Authentication and encryption are the highest security supported by the standard which allows data to be exchanged in a BAN with authentication and encryption.

The standard requires that devices joining the network select one of the three security levels. The security measures defined in the IEEE 802.15.6 standard can be effectively used to protect the BAN communications; however, there are limitations in the standard which makes it unsuitable for practical implementations [42].

8.4.2.1 Securing Communication

The need to secure communications in BAN has resulted in deployment of encryption schemes to protect the data communication. However, due to limited processing capabilities of the IoT devices utilized in eHealth applications, light weight schemes such as elliptic curve cryptography is being widely adopted due to its lower computation overhead [42].

In the case of implantable medical devices (IMD), security mechanisms have been designed to protect the devices from being altered or controlled by unauthorized users. One such method was proposed in [47] referred to as IMDGaurd which utilizes the patient's electrocardiogram (ECG) signals to extract keys for secure communication between implanted and authorized external device. The IMDGaurd avoids using a preconfigured key and can be rekeyed when external device is lost or malfunctioning. Other IMD security measures use a communication shield which jams all the signals received from the IMD device and only permits communication between verified external device or collect data from IMD device and detect malicious communications [48].

Furthermore, network traffic analysis needs to be performed for security monitoring. Implementation of the secure shell (SSH) tunnel and Socks proxy may ensure Internet security [49]. However, encryption of the network traffic may not be able to ensure the security and privacy for the IoT system, as it is possible to extract information from the encrypted network traffic. Therefore, identity is no more hidden as highlighted in [49], as machine learning-based techniques can be used to identify encrypted channels. As a result, the typical encrypted channel may not be enough to ensure secure and reliable information transfer. Authors in [49] suggest further research on the SSH protocol to ensure security by balancing efficiency.

The lack of processing and storage capabilities in IoT devices has resulted in research focus on using smart eHealth gateways. These smart gateways can store, process, and make decisions closer to the patient and reduce the remote communications to the Cloud-based eHealth nodes. Smart gateways can provide preliminary results to the healthcare providers and enable quicker responses in treatment processes. In addition to supporting various IoT communication protocols, these gateways can also use complex encryption techniques to secure the communication between the healthcare providers and the gateway due to higher processing capabilities [50].

In order to protect the intersensor communication between the sensors in BAN from threats such as eavesdropping, message injection, replay, or tampering attacks, a secure encryption scheme is proposed in [51] which uses physiological signal attributes measured from the patient's body. The physiological signal-based key agreement (PSKA) scheme avoids a complex key agreement and key distribution mechanism required in traditional encryption schemes. This enables sensors in BAN to be connected in a plug-and-play manner to allow a secure intersensor communication which can be extended for securing end-to-end communication in eHealth systems [52].

As most eHealth patient monitoring devices have resource constraints, providing end-to-end secure communication is challenging as it requires complex computations for encrypting the messages. Authors in [53] have proposed to offload computation work to third-party entities to reduce the computation load on IoT devices. In the scheme proposed in [53], the devices can perform symmetric encryption with the third parties by randomly splitting the secret key based on the number of third parties involved. The third parties then perform asymmetric encryption with the remote server which then decrypts and reassembles the messages to recover the full message.

8.4.3 Adaptive Security Framework

The eHealth ecosystem has a dynamic nature due to the introduction of new heterogeneous devices and solutions. Hence, the threats to it are also changing and require a flexible security measure that takes into consideration the changing context [54]. The traditional security and privacy solutions rely on a fixed method and all the devices in the system need to adjust accordingly. This can lead to the compromise of the entire system at later point of time when vulnerabilities are found.

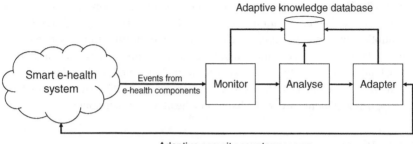

Figure 8.5 Adaptive security steps in protecting eHealth systems. Source: Adapted from Habib and Leister [20].

Hence, the security and privacy solutions deployed in eHealth systems must be adaptive, context-aware, and scalable to protect from various threats.

The adaptive security in eHealth can be applied to monitor and analyze events generated from various components (IoT devices, smart gateways, Fog nodes, Cloud nodes, and HIN) that constitute the eHealth system as shown in Figure 8.5. In addition to events, the context of eHealth components can be included to make context-aware adaptation for securing the system. An adapter module takes intelligent decisions to adapt to the changing environment [20] based on the events generated by the eHealth components. Machine learning-based techniques can be applied in the adapter module to generate intelligent decisions based on the reported events. Previous decisions can be stored in a knowledge database which improves the decision-making process in the adapter module. One such example is an implementation of adaptive authentication [55] which uses channel characteristics and a machine learning-based risk score to develop authentication solution for constrained devices in eHealth system.

A risk-based adaptive framework for IoT-enabled eHealth systems is presented in [54]. The authors first highlight the significance of having a risk management process and a practice in place for eHealth systems. They further studied IoT-enabled eHealth systems, and how they are vulnerable to cyber threats. The lack of energy resources in IoT devices are highlighted as a major hurdle toward deployment of legacy security solutions such as those that involve heavy mathematics and consequent computing resources (e.g. AES-256 bits for data encryption), on resources-constrained IoT devices. The proposed adaptive risk management framework comprises the following modules: adaptive monitoring, analytics and predictive models, and an adaptive model.

The adaptive monitoring module is responsible for IoT sensors, correlation analysis of data and reporting. The analytics and predictive models comprise machine learning, game theory, rule-based reasoning, and context awareness applications to the IoT eHealth data. The adaptive models include decision-making, security and privacy, and actuators for analysis of the IoT data. The proposed framework has implications on applications such as patient monitoring studies, wherein the sensory data as generated from wearable IoT sensors are transmitted to a centralized Cloud storage facility, at which point the analytics module does data processing in order to feed in a decision support system and to facilitate decision-making for supporting patient care in the hands of a healthcare practitioner.

8.4.4 Use Cases

M2M communication allows for machines to communicate with other machines for exchanging data for various application domains. In the context of eHealth, M2M communication would facilitate the communication between various eHealth devices required for patient monitoring such as heart rate and blood pressure monitors and implants, with other centralized entities such as gateways in a patient monitoring neighborhood and centralized Cloud storage facilities. The remote monitoring activities for patients are thus realized through such a setup [56].

M2M devices exchange patient data with a gateway, which is further connected to a core wide area network (WAN) that forward the patient data to M2M servers and other application layer software, for data analytics and decision-making. The security of an M2M platform comprises device and user-level authentication to prevent unknown and malicious entities from communicating with the device, authorization of users in order to prevent escalation of privileges by authentic users accessing M2M devices, management of encryption keys, essential to confirm to encryption standards on M2M devices, and communication networks, thus preventing information leakage that may lead to privacy compromise attacks against sensitive patient data [56].

In [18], the authors have presented an smart eHealth gateway case study, wherein the smart gateway, namely, the UT-GATE, collaborates with sensor nodes, centralized servers, and client devices. The platform facilitates communication of sensory node data as captured from eHealth sensors such as those appended to patients' bodies as part of a BAN, with centralized servers and subsequently with web-based end users. The case study incorporates the implementation of various network topologies, star and mesh, and also comprise the deployment of communication standards including 6LoWPAN, Wi-Fi, and Bluetooth; heterogeneity in standards allows for various sensors to communicate with the centralized servers. Security is provisioned through an Uncomplicated Firewall (UFW) that operates to restrict port and protocol access to the UT-GATE gateway from unauthorized users. In [57], the authors have presented a respiratory rehabilitation use case for eHealth applications, wherein two main scenarios have been studied, namely, ambulatory and home-based. Both these scenarios emulate standard hospital environments. While maintaining the standard process for data fusion from sensors as deployed in an IoT-based eHealth setup, the security of the schemes are ascertained through confidentiality and privacy of data based on Transport Layer Security (TLS), which is supported by the IoT standards of the eHealth system, namely, data distribution service (DDS) and MQTT.

In [58], a Fog-based eHealth platform is presented comprising of smart eyeglasses for heart rate monitoring, Fog node for aggregating the smart class heart rate values, and for conversion into XML format and watermarking, and also a module to upload the processed data onto a centralized eHealth Cloud server. Although the topology of the platform is robust and comprises of various aspects of data fusion and processing, the security of the platform for data at rest, in use as well as at motion, is not addressed in the case study.

Some previous studies have focused on the provisioning of security context frameworks for distributed healthcare systems. In [15], the framework for securing a healthcare system comprises several modules, namely, a security context, an action, and a propagation context. Within each of these modules, there exists a prime source of reference point for information that appertains to health data. The security context comprises an access control list (ACL) and an audit list of actions. This context module is essential in providing only privilege-based access to critical and sensitive patient data. The action module is where a security principal such as the healthcare entity, a smart city device seeking health records (e.g. a governmental agency that is working on patient healthcare data for auditing purposes) carries out an operation on the data that are acquired from the Cloud-based storage systems. The propagation context module is responsible for the exchange procedure definitions for the data that is moving from one entity to another, with potential implications for smart city data movement appertaining to healthcare data.

8.5 Conclusion

The eHealth architectures as presented in this chapter validate the discussion that surrounds the lack of security, and how such a gap in technological adoption can result in unwanted threats posed to the eHealth architectures from the adversary. Patient data are critical in terms of its sensitivity, and their preservation cannot be achieved without the presence of proper security controls and processes. We have highlighted the key challenges appertaining to security of eHealth architectures and have also presented the various techniques and tools that exist in the literature, for enabling security of eHealth architectures.

References

1 Schiza, E.C., Neokleous, K.C., Petkov, N., and Schizas, C.N. (2015). A patient centered electronic health: eHealth system development. *Technology and Health Care* 23 (4): 509–522.

2 Noar, S.M. and Harrington, N.G. (2012). *eHealth Applications: Promising Strategies for Behavior Change*. Routledge.

3 Genes, N., Violante, S., Cetrangol, C. et al. (2018). From smartphone to EHR: a case report on integrating patient-generated health data. *NPJ Digital Medicine* 1 (1): 1–6.

4 Firouzi, F., Farahani, B., Ibrahim, M., and Chakrabarty, K. (2018). Keynote paper: from EDA to IoT eHealth: promises, challenges, and solutions. *IEEE Transactions on Computer-Aided Design of Integrated Circuits and Systems* 37 (12): 2965–2978.

5 Islam, S.M.R., Kwak, D., Kabir, M.H. et al. (2015). The internet of things for health care: a comprehensive survey. *IEEE Access* 3: 678–708.

6 Dash, S., Shakyawar, S.K., Sharma, M., and Kaushik, S. (2019). Big data in healthcare: management, analysis and future prospects. *Journal of Big Data* 6 (1): 1–25.

7 Laplante, P.A. and Laplante, N. (2016). The internet of things in healthcare: potential applications and challenges. *It Professional* 18 (3): 2–4.

8 Yacchirema, D.C., Sarabia-J Come, D., Palau, C.E., and Esteve, M. (2018). A smart system for sleep monitoring by integrating IoT with big data analytics. *IEEE Access* 6: 35988–36001.

9 Garc a-Magari o, I., Lacuesta, R., and Lloret, J. (2018). Agent-based simulation of smart beds with internet-of-things for exploring big data analytics. *IEEE Access* 6: 366–379.

10 Hossain, M.S. and Muhammad, G. (2018). Emotion-aware connected health-care big data towards 5G. *IEEE Internet of Things Journal* 5 (4): 2399–2406.

11 Sundaravadivel, P., Kesavan, K., Kesavan, L. et al. (2018). Smart-log: a deep-learning based automated nutrition monitoring system in the IoT. *IEEE Transactions on Consumer Electronics* 64 (3): 390–398.

12 Davis, J. (2019). Healthcare cyberattacks cost $1.4 million on average in recovery. https://healthitsecurity.com/news/healthcare-cyberattacks-cost-1.4-million-on-average-in-recovery (accessed 25 September 2021).

13 Coventry, L. and Branley, D. (2018). Cybersecurity in healthcare: a narrative review of trends, threats and ways forward. *Maturitas* 113: 48–52. https://doi.org/10.1016/j.maturitas.2018.04.008.

14 Baig, Z.A., Szewczyk, P., Valli, C. et al. (2017). Future challenges for smart cities: cyber-security and digital forensics. *Digital Investigation* 22: 3–13.

15 Sangpetch, O. and Sangpetch, A. (2016). Security context framework for distributed healthcare IoT platform. *International Conference on IoT Technologies for HealthCare*, pp. 71–76. Springer.

16 Medium (2019). IoT in healthcare: are we witnessing a new revolution?. https://medium.com/sciforce/iot-in-healthcare-are-we-witnessing-a-new-revolution-6bb0ecf55991 (accessed 25 September 2021).

17 Lake, D., Milito, R.M.R., Morrow, M., and Vargheese, R. (2014). Internet of things: architectural framework for eHealth security. *Journal of ICT Standardization* 1 (3): 301–328.

18 Rahmani, A.-M., Thanigaivelan, N.K., Gia, T.N. et al. (2015). Smart e-Health gateway: bringing intelligence to internet-of-things based ubiquitous healthcare systems. *2015 12th Annual IEEE Consumer Communications and Networking Conference (CCNC)*, pp. 826–834. IEEE.

19 ISE (2016). Securing hospitals. ise.io/wp-content/uploads/2017/07/securing_hospitals.pdf (accessed 25 September 2021).

20 Habib, K. and Leister, W. (2015). Threats identification for the smart internet of things in eHealth and adaptive security countermeasures. *2015 7th International Conference on New Technologies, Mobility and Security (NTMS)*, pp. 1–5. IEEE.

21 Safavi, S., Meer, A.M., Melanie, E.K.J., and Shukur, Z. (2018). Cyber vulnerabilities on smart healthcare, review and solutions. *2018 Cyber Resilience Conference (CRC)*, pp. 1–5. IEEE.

22 Zeadally, S., Isaac, J.T., and Baig, Z. (2016). Security attacks and solutions in electronic health (e-Health) systems. *Journal of Medical Systems* 40 (12): 263.

23 Ahmed, A., Latif, R., Latif, S. et al. (2018). Malicious insiders attack in IoT based multi-cloud e-Healthcare environment: a systematic literature review. *Multimedia Tools and Applications* 77 (17): 21947–21965.

24 Gordon, W.J., Fairhall, A., and Landman, A. (2017). Threats to information security-public health implications. *New England Journal of Medicine* 377 (8): 707–709.

25 LILY HAY NEWMAN (2017). Medical devices are the next security nightmare. https://www.wired.com/2017/03/medical-devices-next-security-nightmare/ (accessed 25 September 2021).

26 Al-Janabi, S., Al-Shourbaji, I., Shojafar, M., and Shamshirband, S. (2017). Survey of main challenges (security and privacy) in wireless body area networks for healthcare applications. *Egyptian Informatics Journal* 18 (2): 113–122. https://doi.org/10.1016/j.eij.2016.11.001.

27 Koutras, D., Stergiopoulos, G., Dasaklis, T. et al. (2020). Security in IoMT communications: a survey. *Sensors* 20 (17): 4828.

28 Habib, K., Torjusen, A., and Leister, W. (2015). Security analysis of a patient monitoring system for the internet of things in eHealth. *The 7th International Conference on eHealth, Telemedicine, and Social Medicine (eTELEMED)*.

29 Khan, S., Parkinson, S., and Qin, Y. (2017). Fog computing security: a review of current applications and security solutions. *Journal of Cloud Computing* 6 (1): 19.

30 Omotosho, A., Haruna, B.A., and Olaniyi, O.M. (2019). Threat modeling of internet of things health devices. *Journal of Applied Security Research* 14 (1): 106–121.

31 Yaacoub, J.-P.A., Noura, M., Noura, H.N. et al. (2020). Securing internet of medical things systems: limitations, issues and recommendations. *Future Generation Computer Systems* 105: 581–606.

32 Hassija, V., Chamola, V., Bajpai, B.C. et al. (2021). Security issues in implantable medical devices: fact or fiction? *Sustainable Cities and Society* 66: 102552. https://doi.org/10.1016/j.scs.2020.102552.

33 Pycroft, L., Boccard, S.G., Owen, S.L.F. et al. (2016). Brainjacking: implant security issues in invasive neuromodulation. *World Neurosurgery* 92: 454–462.

34 Halperin, D., Heydt-Benjamin, T.S., Ransford, B. et al. (2008). Pacemakers and implantable cardiac defibrillators: software radio attacks and zero-power defenses. *2008 IEEE Symposium on Security and Privacy (sp 2008)*, pp. 129–142. IEEE.

35 Reuters (2016). Johnson & Johnson letter on cyber bug in insulin pump. https://www.reuters.com/article/us-johnson-johnson-cyber-insulin-pumps-t-idUSKCN12414G (accessed 25 September 2021).

36 Storm, D. (2015). MEDJACK: Hackers hijacking medical devices to create backdoors in hospital networks. https://www.computerworld.com/article/2932371/medjack-hackers-hijacking-medical-devices-to-create-backdoors-in-hospital-networks.html (accessed 25 September 2021).

37 Le Bris, A. and El Asri, W. (2016). State of cybersecurity & cyber threats in healthcare organizations. *ESSEC Business School*. https://blogs.harvard.edu/cybersecurity/files/2017/01/risks-and-threats-healthcare-strategic-report.pdf

38 Rifi, N., Rachkidi, E., Agoulmine, N., and Taher, N.C. (2017). Towards using blockchain technology for eHealth data access management. *2017 4th International Conference on Advances in Biomedical Engineering (ICABME)*, pp. 1–4. IEEE.

39 Christidis, K. and Devetsikiotis, M. (2016). Blockchains and smart contracts for the internet of things. *IEEE Access* 4: 2292–2303.

40 Zhang, X., Poslad, S., and Ma, Z. (2018). Block-based access control for blockchain-based electronic medical records (EMRs) query in eHealth. *2018 IEEE Global Communications Conference (GLOBECOM)*, pp. 1–7, December 2018. https://doi.org/10.1109/GLOCOM.2018.8647433.

41 Salahuddin, M.A., Al-Fuqaha, A., Guizani, M. et al. (2017). Softwarization of internet of things infrastructure for secure and smart healthcare. *Computer* 50 (7): 74–79.

42 He, D., Zeadally, S., Kumar, N., and Lee, J. (2017). Anonymous authentication for wireless body area networks with provable security. *IEEE Systems Journal* 11 (4): 2590–2601. https://doi.org/10.1109/JSYST.2016.2544805.

43 Yanambaka, V., Mohanty, S., Kougianos, E. et al. (2019). PMsec: PUF-based energy-efficient authentication of devices in the internet of medical things (IoMT). *2019 IEEE International Symposium on Smart Electronic Systems (iSES)(Formerly iNiS)*, pp. 320–321. IEEE.

44 Shah, S.W., Syed, N.F., Shaghaghi, A. et al. (2020). Towards a lightweight continuous authentication protocol for device-to-device communication.

45 Rasmussen, K.B., Castelluccia, C., Heydt-Benjamin, T.S., and Capkun, S. (2009). Proximity-based access control for implantable medical devices. *Proceedings of the 16th ACM Conference on Computer and Communications Security*, CCS '09, pp. 410–419. New York, NY, USA: Association for Computing Machinery. ISBN 9781605588940. https://doi.org/10.1145/1653662. 1653712.

46 Kwak, K.S., Ullah, S., and Ullah, N. (2010). An overview of IEEE 802.15.6 standard. *2010 3rd International Symposium on Applied Sciences in Biomedical and Communication Technologies (ISABEL 2010)*, pp. 1–6. IEEE.

47 Xu, F., Qin, Z., Tan, C.C. et al. (2011). IMDGuard: Securing implantable medical devices with the external wearable guardian. *2011 Proceedings IEEE INFOCOM*, pp. 1862–1870. IEEE.

48 Zheng, G., Shankaran, R., Orgun, M.A. et al. (2016). Ideas and challenges for securing wireless implantable medical devices: a review. *IEEE Sensors Journal* 17 (3): 562–576.

49 He, D., Ye, R., Chan, S. et al. (2018). Privacy in the internet of things for smart healthcare. *IEEE Communications Magazine* 56 (4): 38–44.

50 Moosavi, S.R., Gia, T.N., Rahmani, A.-M. et al. (2015). Sea: a secure and efficient authentication and authorization architecture for IoT-based healthcare using smart gateways. In: *Procedia Computer Science* (eds. E. Shakshuki), vol. 52, 452–459. Elsevier.

51 Venkatasubramanian, K.K., Banerjee, A., and Gupta, S.K.S. (2009). PSKA: Usable and secure key agreement scheme for body area networks. *IEEE Transactions on Information Technology in Biomedicine* 14 (1): 60–68.

52 Suciu, G., Suciu, V., Martian, A. et al. (2015). Big data, internet of things and cloud convergence–an architecture for secure e-health applications. *Journal of Medical Systems* 39 (11): 141.

53 Abdmeziem, M.R. and Tandjaoui, D. (2015). An end-to-end secure key management protocol for e-health applications. *Computers and Electrical Engineering* 44: 184–197.

54 Abie, H. and Balasingham, I. (2012). Risk-based adaptive security for smart IoT in eHealth. *Proceedings of the 7th International Conference on Body Area Networks*, pp. 269–275.

55 Gebrie, M.T. and Abie, H. (2017). Risk-based adaptive authentication for internet of things in smart home eHealth. *Proceedings of the 11th European Conference on Software Architecture: Companion Proceedings*, pp. 102–108.

56 Fan, Z., Haines, R.J., and Kulkarni, P. (2014). M2M communications for e-Health and smart grid: an industry and standard perspective. *IEEE Wireless Communications* 21 (1): 62–69.

57 Talaminos-Barroso, A., Estudillo-Valderrama, M.A., Roa, L.M. et al. (2016). A machine-to-machine protocol benchmark for eHealth applications–use case: respiratory rehabilitation. *Computer Methods and Programs in Biomedicine* 129: 1–11.

58 Farahani, B., Firouzi, F., Chang, V. et al. (2018). Towards fog-driven IoT eHealth: promises and challenges of IoT in medicine and healthcare. *Future Generation Computer Systems* 78: 659–676.

9

Security and Privacy of Smart Homes: Issues and Solutions

Martin Lundgren[1] and Ali Padyab[2]

[1]Department of Computer Science, Electrical and Space Engineering, Luleå University of Technology, Luleå, Sweden
[2]School of Informatics, University of Skövde, Skövde, Sweden

9.1 Introduction

The current discussion and adoption of new technologies such as Internet of Things (IoT) and smart technologies, like smart homes, have blossomed over the last decade. In short, this new trend can be summarized as information and communication technologies and is increasingly becoming embedded into everyday appliances, such as phones and TV, even refrigerators and thermostats [1]. Interlinking different devices and connecting them to the Internet has sparked new ideas and innovations among technology developers, service providers, and energy utilities alike. Indeed, a recent study on customer adoption of smart home technologies carried out by Accenture estimated that smart home technology "will add 14 trillion dollars of economic value to the global economy by 2030" [2].

The potential benefits of smart home applications are broad, promising that "the smart homes of the future will undoubtedly make our lives much more comfortable than ever" [3]. In the race to provide new innovative solutions, technology developers have applied "smart" to a huge variety of home utilities, sometimes with little focus on the actual end users. Indeed, a user-centric aspect has largely been missing from the discussion which is dominated by technology developers [4, 5]. However, the user-centric aspect plays a vital role in the development of smart homes, since its spread and usage is fundamentally depending on people adopting new technologies into their normal everyday lives. One example that shows the importance of the user-centric aspect can be found in the study made by Accenture, which showed that one of the highest barriers to overcome the adoption of smart home technologies was related to privacy concerns [2]. But what

Security and Privacy in the Internet of Things: Architectures, Techniques, and Applications,
First Edition. Edited by Ali Ismail Awad and Jemal Abawajy.

exactly are these concerns, and what security mechanisms are available to overcome this barrier to protect the end user's privacy?

One of the main difficulties with discussing this topic is that it is not always clear what is meant by either privacy or security solutions in that respect. Therefore, the first step is to investigate the concept of privacy within smart home technology, its importance and the gap in the literature, and then define a framework for the analysis of privacy issues. This chapter contributes to raising our understanding of the security and privacy challenges and solutions that exist within smart homes.

This chapter is organized as follows: In Section 9.2, we will first investigate various dimensions of information security and privacy in order to build a framework to analyze actual or perceived security and privacy issues that can arise from new technologies like smart homes. Section 9.3 presents what security techniques and mechanisms are available to address these. Finally, in Section 9.4, we will discuss what the future might hold in terms of security and privacy of smart homes, followed by Section 9.5 which highlights the contributions of this chapter.

9.2 State-of-the-Art in Smart Homes' Security and Privacy

9.2.1 Smart Home Technologies

Facilitated by IoT, smart home devices have been adopted widely by household owners to fulfill a wide array of functions and needs. The definition of "smart home" could be understood related to its building words smart and home in which its goal is to bring comfort, healthcare, safety, security, and energy conservation to a home [6]. The term "smart homes" is related to home appliances (e.g. TV, security cameras, thermostats, door locks, and refrigerators) which are connected to the Internet [7, 8], and can form other types such as residences (e.g. an apartment, a house, or a unit) equipped with sensors that without any intervention from their inhabitants, automatically collect data from its surroundings and autonomously react based on the scenario it actively infers [9]. The services and features provided through smart homes are vast. Balta-Ozkan et al. [10] characterized services to three broader, overarching categories of safety, energy management, and lifestyle support.

The idea of smart home technologies is to create a network of devices that work in harmony in such a way that specified performance measures are optimized [11]. In this regard, there are different characteristics of these technologies that play an important role for them to function properly. The notion of smart homes is usually taken from a pool of interconnected devices through telecommunications networks, especially the Internet. Some of the devices used in a smart home environment do not have dedicated processing power of their own, but instead collects data which is sent to other units for inferences. Examples of these types are sensors and actuators, which are not necessarily smart on their own but are dependent on another unit, such as a microcontroller, that has the capacity to collectively store and analyze data in order to extract patterns and make decisions [12]. This implies that architectural design and technical characteristics need to be considered when analyzing smart home systems.

From an architectural point of view, a majority of designs are based on cloud computing due to the limited processing power of smart home devices [13]. In such a design, sensors, actuators, and devices communicate via different communication means including wired, wireless, and hybrid to a central server in the cloud, where data is stored, processed, and decisions are made based on the user preferences. Moreover, applications can be made via available data, where services are provided to the users who can get visualizations and control devices via applications [14]. Other architectures include Fog computing and edge computing to ease the CPU power and speed, network transmission load, unacceptable latency, lack of mobility support, and location-awareness on the cloud, at the perimeter of the smart home [15]. In such designs, data can be processed in between the data source and the cloud, before the trimmed down data is sent to the cloud. Examples of integrating smart homes in Fog and edge computing are available in the work of Yi et al. and Okay and Ozdemir [15, 16].

9.2.2 User-Centric Privacy

While the development and deployment of smart homes promise benefits and comfort to our daily lives, they also come with unique security challenges. The realization of smart homes stems from the interconnectivity through everyday devices, pervasive computing that turns "things" into "smart" objects. Looking at this development from the aspect of information security, the model is shifting from an enterprise-centric approach to that of a user-centric perspective [17]. The difference lies in the discomposure of the interconnected devices data and the role of security demands of the intended user.

Compared with a user-centric perspective, an enterprise-centric environment is usually controlled to some extent – be it policy- or technology-wise – as part of a bigger information security program. Security controls (e.g. firewalls, two-factor authentication, and antivirus software) are an essential part of most programs and aim at controlling sensitive information from being disclosed or manipulated – be it in transmission, storage, or when processed [18]. Implementing, managing, and maintaining such security measures have shown to require a great deal of knowledge and experience [19, 20]. However, as we go further toward a user-centric perspective, where the deployment environment is often uncertain and where basic security controls are lacking, security falls with the average user who may lack the required knowledge and experience to understand and secure the technology. The average user is referred here to as the general population, the broader masses of end users and consumers of smart home technologies. But as we will discuss, the responsibility of securing sensitive information cannot fall solely with this audience.

Recent privacy and data protection laws have thus called for more involvement with the end user to have a say in protecting their private information (e.g. General Data Protection Regulation (GDPR)). This has commonly resulted in device policies, developed to enable end users to determine who, when, and for what purpose their information should be shared. Although such policies seem like a solution, it has been shown to come with several shortcomings [21–23]. For example, such an approach assumes that the end users fully understand these privacy policies and encourages end users to demand their rights from the service providers. However, problems arise when end users are not fully aware of possible security risks related to the technology, and the direct or indirect consequences associated with personal information disclosure as the result of a data breach [24]. Article 4 (12) of the GDPR define such a data breach as "a breach of security leading to the accidental or unlawful destruction, loss, alteration, unauthorized disclosure of, or access to, personal data transmitted, stored or otherwise processed" [25] This definition not only touches upon the risk of a data breach but also includes the security of data and the state in which it is in. The latter can have severe consequences on how privacy is perceived, and how new technologies are ultimately adopted.

9.2.3 Consequences of Data Breaches

The topic of data breaches is commonly discussed within information security. There are many different sources that can cause a data breach, such as incidents, dedicated attacks, improper disposal, loss, theft, and unauthorized access or disclosure of information [26]. In order to secure an organizations environment, there is a need to not only identify which possible breaches could occur but also to determine the possible impact or damage should some information or data be wrongfully disclosed, manipulated, or turn inaccessible due to a breach. Determining the impact can help prioritize which potential breaches to manage first. There are several widely used aspects and dimensions to estimate how severely an organization could be affected. For example, for a service provider, we can look at the loss of revenue as a result of the breach, loss of intellectual property [18], or damage to brand or reputation [27].

The sources of data breaches are not as clear when it comes to personal data. For example, a personal data breach can take many shapes and forms. A device's privacy policy might allow it to share the information it collects with the company that makes the device, or a third party for that matter. Or there might be a security breach into the device, and someone is actively stealing or observing the information. Neither of these cases fall under the aforementioned dimensions of a typical information security data breach. Instead, consequences of data breaches for an end user include psychological, social, career-related, physical, resource-related, prosecution-related, and freedom-related [28]. Privacy as a dimension to data breaches within information security has long been lagging behind [29], yet privacy has shown to be an affecting factor in adopting smart home technology [2].

9.2.4 Dimensions of Privacy Concerns

A measurable value is usually associated with the consequences of data breaches. However, consequences of privacy concerns might be better suited for this particular issue, since privacy is relative to the person and how he or she perceives themselves in a particular context [30, 31]. The distinction is subtle, but important. Looking into research that has been carried out within the context of smart homes, issues related to data breaches have been described in different terms than loss of value. One such example can be found in a study from 2008 that looked at senior's willingness to adapt to smart home technology [32]. The study found that privacy concerns could be defined as simply as a desire to be alone, to control the information shared with others, to control access to one's personal property, and to protect oneself from identity theft [32]. But in the study, little was spoken about monetary or otherwise measurable loss. Instead, one can draw from Barhamgi et al.'s [22] study on user-centric privacy and concerns that summarize privacy concerns into

- **Privacy of personal information** (i.e. data generated by or about the user);
- **Privacy of personal behavior** (i.e. the person, or the integrity of his or her body); and
- **Privacy of personal communications** (i.e. conversations, texts, and chats).

Each of these three dimensions provide their own subdimensions to privacy and can be considered when selecting and implementing security controls designed to protect end user's data. These dimensions of privacy concerns also illuminate the end user's privacy, as each can easily be expanded to include distinct examples and scenarios. Expanding these dimensions into the context of smart homes bring forth the importance of privacy considerations in the design of smart home systems.

Regarding the privacy of the *personal information,* for example breaches such as identity theft and personal information hijacking in a smart home environment [33, 34], as well as leakage of camera feed installed in the home due to its misconfiguration [35], are among the various personal information privacy consequences.

The *privacy of personal behavior* is related to the psychology and personality traits of the inhabitants. Such information has strategic value to the companies, service providers, governments, and hackers for various purposes such as targeted advertisements, behavioral inferencing, and for nefarious purposes such as discrimination or to pursue a political agenda. One recent example concerns data analytic firm, Cambridge Analytica, and its use of 87 million people's Facebook data for behavioral data mining and crafting personalized micro-ads, targeting voters with the aim to influence their votes in the 2016 US presidential election [36]. This practice is viable to all sorts of data gathered via smart home devices, and, due to the rich contextual information hidden within the data, it makes a valuable source for the attacker.

With respect to the *privacy of personal communications*, information communicated should only be made available to the intended audiences. This dimension has two aspects. First, information collected and processed by the device vendor, service provider, or third parties could be disclosed beyond user's awareness. Smart home devices make personal communication, in the most intimate areas of a person's life (i.e. home), vulnerable, and there are many examples of physical surveillance through smart devices. In 2017, for example a warning was issued by the Federal Network Agency (Bundesnetzagentur) in Germany for parents to destroy a particular brand of a talking doll after it was revealed its smart technology could disclose personal information [37]. Or when it came to light in 2015 that a certain brand of smart TV's captured and transmitted any spoken word nearby to a third party through its voice activation feature [38]. Second, communication has two sides: the sender and the receiver. The receiver of the information infers the information and associates it with the sender. In this regard, any compromise to the communication channel is perceived as a violation of privacy, since information posed from an unreliable source could paint a wrongful picture of the sender. Proofpoint reports that more than 750,000 phishing and SPAM e-mails launched from "Thingbots" including televisions, fridge, thermostats, and smart locks [39]. Such a large-scale attack implies that the owners of smart home devices could be regarded as spammers and shows that the security of such devices is linked to the privacy of personal communication.

9.2.5 Consequences of Information Security

Security challenges that come with smart home technology are different, and at the same time similar to the security of any IT environment [40, 41]. As with IT, smart home devices suffer from similar challenges regarding how information is transmitted, stored, and processed, but can therefore also be analyzed from a similar security goals as any IT environment. For many years, the CIA-triad (Confidentiality, Integrity, and Availability) has been a guiding principle for security measures. Each of which brings important aspects to information security to be considered.

- **Confidentiality** (i.e. information only being accessible to those authorized and with sufficient privileges);
- **Integrity** (i.e. information being complete, whole, and uncorrupted); and
- **Availability** (i.e. information being accessible in a usable format when needed).

A lot of the threats that target smart home technology has consequences for end users' privacy; therefore, reminds us about the challenges that has long been entangled with the IT domain. Below, we explain each aspect in relation to the smart home:

Confidentiality is an important aspect of many smart devices, since many devices are usually connected to some form of network and could be transmitting confidential data [42]. Hence, the communication between the device and an end-point, e.g. the cloud, must follow a protocol that allows for a secure connection not to leak or otherwise be tricked into revealing confidential data [43]. One such example is to interrupt a Bluetooth pairing communication between two devices, masquerading as the real receiver or sender, thus either sending false data or reading confidential, possibly sensitive, data [42]. Indeed, when it comes to network security, smart homes are no different from other connected IT devices in terms of possible attacks, such as network packet sniffers, password attacks, IP spoofing, man-in-the-middle (MITM) attacks, and distribution of sensitive internal information to external sources [44].

Similarly, *integrity* is also important to smart home devices. For example, many firmware updates are not digitally signed and still rely on Trivial File Transfer Protocol (TFTP) servers. The lack of proper integrity checks could be easily exploited by a breach to the local network. By using Address Resolution Protocol (ARP) spoofing, for example an attacker could have smart home devices connect to a nefarious TFTP server and feed them manipulated firmware updates [42]. Another example is the modification of packets in a smart grid where the attacker alters the contents of messages, either by increasing the amount of energy to be discharged from houses to the grid or by decreasing the amount of energy to be absorbed from the grid. If massive, such an attack could cause the grid to be unstable [45].

Availability is certainly also an important security aspect for smart home devices. During recent years, there have been plenty of examples where smart home devices, as well as other types of IoT devices, have been breached by attackers and included as part of a bigger collection of breached devices (often referred to as a zombie network or botnet) in order to launch Distributed Denial of Service Attacks (DDoS), as was the case of the Mirai botnet [46]. The consequences could lead to device malfunction, bring about financial losses due to a rise in electric rate, and could even lead to risk of life due to device overload that could trigger a fire [47].

Recognizing the similarities between smart home devices and other IT environments, the discussion around security challenges in smart homes can borrow a lot from the field of information security. However, the discussion must further the recognition of privacy as a potential impact. Bridging the CIA triad with dimensions of privacy might give a better understanding of how information security can aid potential solutions toward an increased smart home adoption.

9.2.6 A Framework for Security and Privacy Concerns

Within the realm of privacy, and in particular within the domain of smart home technology, the CIA triad falls short with respect to the information states. As discussed in previous examples, the information state – e.g. where and how information is stored, if it is in rest or in motion, and how the information is used and processed – can play a role in how privacy is perceived by the end user. As such, these aspects of the information state can act as additional dimensions to the CIA-triad, not only in terms of aiding in security planning and evaluation but also in terms of assessing consequences of privacy concerns as well.

Including dimensions of information state onto the CIA triad is nothing new and has previously been described as a three-dimensional cube, known as the McCumber cube [48]. However, the concept of the McCumber cube takes it one step further and expands the traditional CIA triad not only by adding three dimension of information states to it but also three dimensions of countermeasures to be considered, respectively. The three information states are described as follows:

- **Transmission** (i.e. information being sent to one or more receivers);
- **Storage** (i.e. information in rest, being saved and gathered); and
- **Processing** (i.e. information in use, handled, prepared, or refined).

The McCumber cube's three dimensions of countermeasures are the following: technological, policy, or educational [49]. Put together, the McCumber cube can help security professionals analyze a given environment to evaluate and plan for what countermeasures to implement in order to control the environment from or mitigate potential impacts of a data breach. As such, it is well suited for enterprise-centric environments. However, in light of the discussions posed in this chapter, we propose a slight redesign of the McCumber cube, shifting it toward a user-centric privacy perspective instead. In the new model (see Figure 9.1), the three dimension of countermeasures in the McCumber cube are removed, and in their place, the three aspects of privacy concerns discussed earlier are added. Consequences of privacy can be considered as one of the three: personal communication, personal information, and personal behavior. The adjustment makes it possible for security professionals and developers to consider security impacts not in terms of security requirements (i.e. countermeasures), but in terms of consequences of resulting privacy concerns as perceived by end users.

Figure 9.1 A user-centric privacy model to capture privacy concerns in relation with information security.

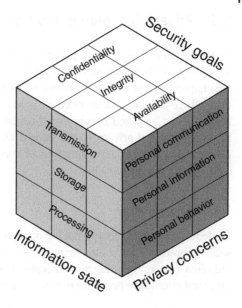

Analyzing smart home technology using the proposed, multidimensional approach can provide a bridged view between security and privacy. For example, considering the security goal "confidentiality" alone of personal information would be rather broad, and not clear if it is in reference to how a smart home device is storing, processing, or transmitting information somewhere. The second dimension, that of information state, leads the discussion into how and where the information resides, is handled, and presented. Lastly, it is important to recognize that the information state is directly related to the consequence of privacy concerns. A breach of an individual's privacy could be the result of compromised personal communication and thereby fall under the privacy concern "personal information." In addition to facilitating a more nuanced view of privacy in smart home technology (or any technology for that matter), it could also give an idea of how to categorize security controls in relation to privacy concerns.

The cube can also be read from the other direction around, starting from privacy concerns. Since privacy is subjective, a perceived breach of privacy can cause a user not to adopt a specific technology. In this sense, a person might feel that, e.g. a particular device affect his or her "personal behavior" because of uncertainty about whether or not the location data is "stored" outside the premises of the end user. Turning to the device vendor, the user can now ask how the device maintains different "information states," and for what purpose. Each of these questions can then be followed-up by the three aspects of the CIA triad and how the different security goals are reached.

9.3 Privacy Techniques and Mechanisms

In this section, the authors present a literature review and seek to identify available security solutions with regard to related privacy concerns. The rest of the chapter aims at a critical analysis of related issues and solutions with respect to the three-dimensional user-centric privacy model. Figure 9.2 shows a concept matrix of the analyzed papers and the extent to which the solutions cover the three-dimensional privacy model. Moreover, each solution is discussed in more detail, followed by the presentation of some real-world cases of smart home security and privacy breaches.

9.3.1 Cryptography

Inadequate network security is a big issue in smart home technologies, and most security protocols for computer networks cannot be implemented because they are often too computationally expensive [50–52]. Some authors have therefore proposed alternative protocols to ease the computational workload. For example,

Privacy techniques and mechanisms	Security goals			Information states			Privacy concerns		
	Confidentiality	Integrity	Availability	Transmission	Storage	Processing	Personal communication	Personal information	Personal behavior
Zeng et al. (2017) [90]	•			•			•	•	•
Apthorpe et al. (2017) [4]	•			•					•
Copos et al. (2016) [18]	•			•					•
Notora et al. (2014) [46]	•								
Chakravorty et al. (2013) [14]	•			•	•	•			
Chifor et al. (2018) [15]	•			•		•			
Li (2013) [35]	•	•		•					
Salami et al. (2016) [55]	•			•					
Kumar et al. (2017) [32]	•	•		•			•	•	
Seigneur et al. (2003) [60]	•			•					
Nobakht et al. (2016) [44]	•		•	•					
Cho et al. (2007) [16]	•			•					
Jose et al. (2016) [28]	•			•				•	
Young-Pil et al. (2015) [86]	•			•				•	
Mantoro et al. (2014) [38]	•	•		•					
Sivaraman et al. (2015) [63]	•		•	•					
Kang et al. (2017) [30]	•	•	•	•					
Bagüés et al. (2007) [6]	•			•			•	•	
Fabian and Feldhaus (2014) [22]	•			•		•		•	
Hoang and Pishva (2015) [25]	•			•				•	

Figure 9.2 An overview of available solutions with respect to the user-centric privacy model.

the work of Salami et al. [51] and Li [50] proposed lightweight key establishment protocols. Li [50] based their key establishment protocol on Elliptic Curve Cryptography (ECC) and concluded that lightweight, secure, cryptographic protocols are indeed possible for smart home devices. Salami et al. [51] based their key management scheme on identity-based encryption and stateful Diffie-Hellman (DH). Another example of meeting this challenge can be found in the work of Kim et al. [52] who suggest a dynamic and energy aware authentication scheme for smart home devices. Whereas authors such as Mantoro et al. [53] propose a security model that includes powerful and low-power consumption symmetric block cipher: AES256, Ephemeral DH Key Exchange to facilitate the key management between the central hub and smart phone.

Challenges concerning computational power are not limited to two-way functions, where data can be scrambled in such a way that it can be unscrambled later. Similar issues exist with one-way functions too. This raises some obvious issues with message integrity in many protocols. Message integrity is therefore often assured via some hash-based schema. However, some smart home devices have limited processing ability, furthermore, sensors are often powered with a batteries that further limits their ability to create hash messages. In this regard, RC4-based hash function could make it more efficient for ultra-low power devices [54], as proposed by Mantoro et al. [53], in which the smart home is provided with the security that is implemented in a smart phone linked to the central hub. Another solution is to outsource the hashing of personal information to a dedicated service located externally. In the work of Chakravorty et al. [55], for example such a solution is proposed. In their study, personally identifiable data are stored outside of the local premises not only as hash values but also as an encrypted dictionary with hashed and actual values of all unique sets of identifiers to enable information processing and reidentification of the results.

Real-World Cases – Cryptography

The ubiquitous nature of smart home devices and their wireless connectivity makes them vulnerable to different man-in-the-middle and Wi-Fi access point attacks. Different examples of such attacks have been reported by independent ethical hacking groups and researchers. One example is a discovery of a man-in-the-middle (MITM) vulnerability in a Samsung smart refrigerator (model number RF28HMELBSR) that can be exploited to gain Google account login credentials. In the workshop at DefCon 23's IoT Village in 2015, researchers embarked on a hacking challenge of Samsung smart fridge which revealed that, although the fridge implements SSL, it failed to validate SSL certificates. This gave the opportunity to the attacker to send a fake certificate to the fridge (via de-authentication and fake Wi-Fi access point attack) to monitor activity for the username and password used to link the refrigerator to Gmail Calendar [56].

Many smart home devices are managed via a mobile application in which the traffic between the two is prone to different spoofing and MITM attacks. Researchers at the KTH University were able to perform a MITM attack against robot vacuum cleaner Ironpie m6 on the mobile application with client-side manipulation [57]. The researchers successfully set up a "mitmproxy" (via ARP spoofing) [58] and bypassed the encryption used in HTTPS to view the contents of the packets. The attack showed that it was possible to capture HTTPS login requests and read the e-mail (Username) and demonstrated how vulnerable HTTPS is to MITM attacks.

9.3.2 Access Control

Many smart home devices allow the end user to control it in some way, e.g. to reconfigure the settings. This, of course, has also raised a lot of issues with how to manage access control to these devices, and for what purpose. In the study conducted by Chifor et al. [59], for example the authors discussed the implications of having smart home devices connected to a cloud-based solution to gain access and process data. Chifor et al. [59] propose a "Fast Identity Online" (FIDO) based solution for a federated authorization mechanism that could be controlled by the end user's smart phone in order to manage access control of the IoT device. Other solutions have also been proposed to overcome similar challenges. In their security framework, Kang et al. [47] created an access control list to make sure that existing modules or a newly added module are properly authenticated from the authentication of mandatory access control framework on kernel level. In another study, access control lists were implemented at the edge of the perimeter (e.g. firewall or ISP) instead of the individual devices or the central hub [35]. Notora et al. [60] similarly suggested access control lists, but at the Internet Service Providers (ISP) level, to rule who had access to what device, and thus limiting possible vulnerable devices to allow external connections to be made to the device. While authors such as Jose et al. [61] propose using a central web-access point to administrate the security devices. The web-access point contains fingerprints of endpoints allowed to connect to smart home devices, based on JavaScript, Flash, and Geo-location.

Seigneur et al. [62] takes the discussion one step further and argued that since many smart home technology users are technology-unaware, they often lack the necessary skills to administrate the device. Instead, Seigneur et al. [62] proposed a security framework where access control rights evolve automatically over time, according to the human notion of trust, thus minimizing security configuration on behalf of the end user. The consequence of trusting the end users with security configurations is exemplified in the work of Nobakht et al. [63]. In their study, Nobakht et al. [63] proposed using a host-based intrusion detection and mitigation framework (IoT-IDM) for smart home devices to overcome the issues of inadequate security configurations. Using machine learning techniques to detect compromised hosts, IoT-IDM can automatically generate and deploy appropriate access policies.

Real-World Cases – Access Control

Hijacking attacks were performed on Google Chromecast by Bishopfox Labs [64]. While Chromecast uses an SSL service, the device functioned over clear text making it possible to obtain Wi-Fi access point Service Set Identifier (SSID). Next, the attacker performs a "de-authentication" attack that disconnects the Chromecast from the Wi-Fi network and forced the Chromecast to go into the pairing mode. The attacker could then force the Chromecast to connect to a rouge access point (with the same Wi-Fi SSID) and stream arbitrary content to the user's screen and, for example playing a voice asking home assistance devices such as Alexa to perform certain activities like controlling home climate temperature [65].

In early March 2021, hackers managed to breach roughly 150,000 Internet-connected surveillance cameras deployed over the Internet. The breach gave the attackers access to live and archived video feed from a series of organizations, including facilities such as hospitals, schools, police departments, and prisons [66]. The San Mateo, California-based manufacturer, Verkada Inc., came out with an official statement on March 11, confirming the attack [67]. The hackers were able to authenticate themselves to the implemented access control and thereby gain full access to such a vast amount of security cameras due to a compromised super-user account that could be found publicly exposed on the Internet. The credentials for the super-user acted as a skeleton key that helped pave the way for the attackers [68].

9.3.3 Privacy Policy

In the context of smart homes, a privacy proxy can act on behalf of the user as a central privacy enforcement point for all privacy-relevant accesses going from and to devices in which the context of access is at the discretion of the user. Sentry@HOME is a context-aware framework which is based on a set of privacy policies defined by the end user. The framework aims to help control the dissemination of data within the context-aware service interaction chain [69]. However, Cho et al. [70] recognized in their study that few context-aware solutions scale and proposed an integrated preservation system by using a middleware to overcome the challenge. Zeng et al. [71] took this one step further and stressed that security and privacy concerns must be deliberately designed to include multiple users, and not only the device owner, since different people around the device can just as easily be monitored without being aware or having given consent.

Real-World Cases – Privacy Policy

Liberatore and Neil Levine [72] concluded that encryption is not enough to protect user privacy since the analysis of packet size and frequency of encrypted network between an IoT device and the cloud is enough to reveal device-level activity. Other researchers have shown that centralized nature of most smart home platforms makes them susceptible to privacy breaches. In a systematic security evaluation of two popular smart home platforms, Google's Nest platform and Philips Hue, the researchers demonstrated sensor blinding and state confusion attacks via the manipulation of state variables in a centralized data store [73]. The attacks showed that the devices compatible with the abovementioned platforms give access to third-part apps to access and manipulate permission map without the user's consent. This is compounded by the fact that users do not get alerts when such changes are made.

In another study, researchers demonstrated how private sensitive information, such as alarm status, color, brightness, and hue, in Philips Hue could be captured via Wireshark of GET request/response exchanged between the bridge and app, all in plain text [74]. Such information could be used by an adversary to gain insights into the current state of affairs inside user's house. In the same study, the authors raised concerns about the daily transfer of 20 KB of data to the log servers from the Nest's smoke detector device. Although the data packet is encrypted, and therefore the contents were unreadable, the authors argued that preventing the data transfer did not jeopardize Nest's ability of notifying user in emergency. This raises the question of the nature of the log transfer worrisome due to the privacy issues [74].

9.3.4 Anonymity

Anonymity is to conceal information related to the initiator, target, as well as query unlinkability. In order to enhance the anonymity of end users when communicating with service providers (e.g. a smart fridge orders food from a local retail store), Fabian and Feldhaus [75], and Hoang and Pishva [76], suggested a privacy-preserving peer-to-peer (P2P) infrastructure using multiple encryption layers (similar to "onions" in TOR [77]). Some studies have illustrated the reality of this concern [78, 79]. For example, a study made by Apthorpe et al. [78] analyzed four commercially available smart home devices and found that a passive network observer could infer user behaviors – even though the traffic was encrypted. The authors argue that policies could help in this matter to regulate how ISP are allowed to inspect the traffic [78], since that would not impact the performance in the same way as other anonymizing solutions, e.g. TOR. Copos et al. [79] similarly argued that a proxy that directed network traffic could act as a shield against some information leakage. However, device anonymity and unlinkability are real challenges that are not easily tackled. In their work, Kumar et al. [33] developed desirable properties for securing smart home devices in a secure, anonymous way, which they presented in their Anonymous Secure Framework (ASF) to provide device anonymity and unlinkability.

Real-World Cases – Anonymity

Smart home devices have over the years suffered from various bugs and mis-configurations that have left end users' information available to attackers and the end users' privacy exposed. In June 2018, for example Tripwire reported a bug in Google's artificial intelligence Home speaker and streaming device Chromecast, that could reveal an end user's precise physical location to an accuracy of around 10 m [80]. Being able to extract end users' GEO-location could among other things, as the researcher from Tripwire noted, lead to more effective blackmail or extortion campaigns [81], or used as a means for harassment, monitoring, revenge, and control [82].

A year later, in July 2019, about two billion user logs from various smart home devices were found to be leaked due to a misconfigured server from a Chinese smart home solutions provider called Orvibo. Researchers at vpnMentor discovered the server accessible from the Internet, with no password protection, in the middle of June [83]. Among the leaked logs were detailed information on user IDs, family names, e-mail addresses, hashed (albeit not salted) password, smart home device details, GEO-location data, IP addresses, as well as account reset codes [84]. VpnMentor reported that the details could allow attackers to access the video feed from one of Orvibo's smart cameras, unlock smart doors, and gain information about end users through their recorded schedule accessible through a smart mirror [83]. The leaked logs resulted in an abundance of identifying information about the smart home end users.

9.3.5 UI/UX, User Awareness, and Control

Zeng et al. [71] studied dimensions of privacy concerns touching upon personal behavior, information, and communication. They found that concerns related to these dimensions of privacy were mostly perceived by users with a more advanced understanding of various security threats. Therefore, the study recommended development of user interface and experience (UI/UX) not only to raise the threat awareness, include reputation systems for smart devices, but also to consciously design smart home devices for multiple users with varying preferences, as with techniques to address privacy and security issues in smart home devices. Seigneur et al. [62] stresses a related point that much of the problem lies in the fact that many home users cannot be considered as skilled administrators, but are often technology-unaware users. In their chapter, Seigneur et al. [62] discuss the implications of putting security in a technology-unaware context and recognize that trust plays an important part.

Real World Cases – UI/UX, User Awareness, and Control

A wide array of smart home devices, such as Samsung SmartThings, Amazon Echo, Google Home, Phillips Hue lights, cameras, and thermostats, allow end users to program their behavior to automate certain applications in daily life [85]. For example, end users can create simple conditional rules for when to turn the light on, or to receive a notification when a door is locked or unlocked. However, studies, like the work of Nandi and Ernst [86] and Milijana et al. [87], have found that end users sometimes make errors when configuring their smart devices, like forgetting to program parts of the intended behavior to misunderstanding the logic flow. The result of which could lead to serious security or privacy concerns, such as incorrectly locking doors, cause fire hazards, or lead to privacy risks due to misconfigured smart cameras [85–87]. Furthermore, a study by Palekar et al. [85] in 2019 found that many smart home device platforms either do not prevent such end users' errors or prevent them accidentally due to feature limitations.

One example of misconfigured smart home devices was found by researchers at Avast [88]. As end users install more and more smart home devices, they sometimes need to be orchestrated from a single point. To overcome the issue of different protocols used by different smart home devices, Message Queuing Telemetry Transport (MQTT) servers can be setup to interconnect the devices. However, in 2018, researchers at Avast [88] found more than 49,000 MQTT servers publicly visible on the Internet – 32,000 of which had no password protection. The exposed devices were believed to have been the result of misconfigured MQTT servers caused by end users when setting up their smart homes [89].

9.4 Toward Future Solutions

Most of the papers analyzed in this chapter elaborate little on the topic of consequences of privacy. Some, like Salami et al. [51] to name one, mention the importance of encrypted network traffic to protect user's privacy. However, little is mentioned on the nature of privacy itself. Therefore, we know little about if the intention is to simply secure, for example the data transmission (which has also shown to be the most commonly addressed issue) from the hackers or to protect the data from anyone without clear user approval. This discussion is important since the mere feeling of violated privacy might affect the user's behavior and adoption of new technology. One such example is the discussion posed by Zeng et al. [71] who recognized that not all users of smart home technology are capable of realizing the threat model, but which can be met by, for example more research into device's user interfaces and experiences.

The discussion on trust could be a dimension to further explore. In the technology dimension of UI/UX, trust was brought up as it relates to the perception of security and privacy. Although UI/UX is not a technology that strengthens either privacy nor security in itself, it is what enables a user to make informed decisions and experience the effect thereof. In this sense, a system might be very well designed both from a security and privacy perspective, but still be rejected by the user if he or she, for one reason or another, does not trust the device.

Interestingly, few of the papers analyzed addressed data when processed and stored. Both storage and processing come, however, with unique characteristics regarding privacy concerns. For example, processing is an interesting aspect since, even though data has been transmitted in a secure way, it is processed in a way to embrace the end user's privacy, and for what purpose? There are several examples of data processing risks, especially when using a physical co-residency in cloud computing, which has shown to introduce several possible side-channel attacks to steal data [90]. Similarly, data storage is also an issue, which carries several examples of exposed data, such as when roughly four million Time Warner Cable customers were exposed due to an improperly configured AWS Simple Storage Service (S3) [91]. Or availability of the information which can get lost forever even in well-structured cloud environments such as when 0.07% of data stored in Amazon Elastic Block Storage volumes got irretrievably lost [91].

9.5 Conclusion

In this chapter, a new three-dimensional model is proposed that contributes to the future research by extending the traditional CIA-triad from an enterprise-centric perspective to a user-centric privacy concerns perspective that can help analyze smart home technology. While many articles do bring up privacy as an important part of securing smart home technologies, they do not often elaborate further on what the violation of privacy involves. The model proposed herein is an attempt to concertize this discussion to further the discussion on privacy concerns and better help identify security requirements that end users of smart home technology can relate to. Despite the smart home literature which was reviewed in this chapter, this overview is far from complete. Moreover, bridging the gap between end user's privacy/security concerns and the development of smart home devices needs a collaboration between researchers from various fields such as computer science, UX, legal, information privacy, and social sciences. We believe that our proposed three-dimensional model can facilitate such collaboration. Our future research aims to employ the framework as an evaluation approach to find out if discrepancies exist between end user's privacy concerns and actual use of smart home devices. In this regard, our next step would be to integrate the model into the design and implementation of such devices.

References

1 Samaila, M.G., Neto, M., Fernandes, D.A.B. et al. (2018). Challenges of securing internet of things devices: a survey. *Security and Privacy* 1 (2): e20. https://doi.org/10.1002/spy2.20.

2 Venturini, F. and Novak, M. (2017). https://www.accenture.com/t20170303T051308Z__w__/us-en/_acnmedia/Accenture/Conversion-Assets/DotCom/Documents/Global/PDF/Dualpub_26/Accenture-The-Race-to-the-Smart-Home.pdf (accessed 24 September 2021).

3 Lin, Y.-J., Latchman, H.A., Lee, M., and Katar, S. (2002). A power line communication network infrastructure for the smart home. *IEEE Wireless Communications* 9 (6): 104–111. https://doi.org/10.1109/MWC.2002.1160088.

4 Solaimani, S., Bouwman, H., and Baken, N. (2011). The smart home landscape: a qualitative meta-analysis. In: *Toward Useful Services for Elderly and People with Disabilities* (ed. B. Abdulrazak, S. Giroux, B. Bouchard et al.), 192–199. Berlin, Heidelberg: Springer-Verlag. ISBN 978-3-642-21535-3.

5 Wilson, C., Hargreaves, T., and Hauxwell-Baldwin, R. (2015). Smart homes and their users: a systematic analysis and key challenges. *Personal and Ubiquitous Computing* 19 (2): 463–476.

6 Alam, M.R., Reaz, M.B.I., and Ali, M.A.M. (2012-11). A review of smart homes-past, present, and future. *IEEE Transactions on Systems, Man, and Cybernetics Part C: Applications and Reviews* 42 (6): 1190–1203. ISSN https://doi.org/10.1109/TSMCC.2012.2189204.

7 Shuhaiber, A. and Mashal, I. (2019). Understanding users' acceptance of smart homes. *Technology in Society* 58: 101110. https://doi.org/10.1016/j.techsoc.2019.01.003.

8 Wu, C-L., Liao, C.-F., and Fu, L.-C. (2007). Service-oriented smart-home architecture based on OSGi and mobile-agent technology. *IEEE Transactions* on *Systems, Man, and Cybernetics, Part C: Applications and Reviews* 37 (2): 193–205. https://doi.org/10.1109/TSMCC.2006.886997.

9 Robles, R.J., Kim, T.-h., Cook, D., and Das, S. (2010). A review on security in smart home development. *International Journal of Advanced Science and Technology* 15. 13–22.

10 Balta-Ozkan, N., Davidson, R., Bicket, M., and Whitmarsh, L. (2013). Social barriers to the adoption of smart homes. *Energy Policy* 63: 363–374. https://doi.org/10.1016/j.enpol.2013.08.043.

11 Cook, D. and Das, S.K. (2004). *Smart Environments: Technology, Protocols, and Applications*. Wiley. ISBN 978-0-471-68658-3.

12 Soliman, M., Abiodun, T., Hamouda, T. et al. (2013). Smart home: integrating internet of things with web services and cloud computing. *2013 IEEE 5th International Conference on Cloud Computing Technology and Science*, Volume 2, pp. 317–320. https://doi.org/10.1109/CloudCom.2013.155.

13 Mocrii, D., Chen, Y., and Musilek, P. (2018). IoT-based smart homes: a review of system architecture, software, communications, privacy and security. *Internet of Things* 1–2: 81–98. https://doi.org/10.1016/j.iot.2018.08.009.

14 Zhou, J., Leppanen, T., Harjula, E. et al. (2013). CloudThings: A common architecture for integrating the internet of things with cloud computing. *Proceedings of the 2013 IEEE 17th International Conference on Computer Supported Cooperative Work in Design (CSCWD)*, pp. 651–657. https://doi.org/10.1109/CSCWD.2013.6581037.

15 Yi, S., Hao, Z., Qin, Z., and Li, Q. (2015). Fog computing: platform and applications. *2015 Third IEEE Workshop on Hot Topics in Web Systems and Technologies (HotWeb)*, pp. 73–78. https://doi.org/10.1109/HotWeb.2015.22.

16 Okay, F.Y. and Ozdemir, S. (2016). A fog computing based smart grid model. *2016 International Symposium on Networks, Computers and Communications (ISNCC)*, pp. 1–6. https://doi.org/10.1109/ISNCC.2016.7746062.

17 Skarmeta, A., Hernández-Ramos, J.L., and Martinez, J.A. (2019). User-centric privacy. In: *Internet of Things Security and Data Protection* (ed. S. Ziegler), 191–209. Cham: Springer International Publishing. ISBN 978-3-030-04983-6 978-3-030-04984-3. https://doi.org/10.1007/978-3-030-04984-3_13.

18 Whitman, M.E. and Mattord, H.J. (2014). *Management of Information Security*. Boston, MA: Cengage Learning.

19 Shedden, P., Smith, W., and Ahmad, A. (2010). Information security risk assessment: towards a business practice perspective. *8th Australian Information Security Mangement Conference*.

20 Wangen, G. (2017). Information security risk assessment: a method comparison. *Computer* 50 (4): 52–61.

21 Aïmeur, E. and Lafond, M. (2013). The scourge of internet personal data collection. *2013 International Conference on Availability, Reliability and Security*, pp. 821–828. https://doi.org/10.1109/ARES.2013.110.

22 Barhamgi, M., Perera, C., Ghedira, C., and Benslimane, D. (2018). User-centric privacy engineering for the internet of things. *IEEE Cloud Computing* 5 (5): 47–57. https://doi.org/10.1109/MCC.2018.053711666.

23 Xu, F., He, J., Wu, X., and Xu, J. (2010). A user-centric privacy access control model. *2010 2nd International Symposium on Information Engineering and Electronic Commerce*, pp. 1–4. https://doi.org/10.1109/IEEC.2010.5533251.

24 Padyab, A., Päivärinta, T., Ståhlbröst, A., and Bergvall-Kåreborn, B. (2019). Awareness of indirect information disclosure on social network sites. *Social Media + Society* 5 (2). https://doi.org/10.1177/2056305118824199.

25 GDPR (2016). European parliament and the council of the European union. https://ec.europa.eu/info/law/law-topic/data-protection_en (accessed 24 September 2021).

26 Williams, P.A.H. and Hossack, E. (2013). It will never happen to us: the likelihood and impact of privacy breaches on health data in australia. *HIC*, pp. 155–161.

27 Syed, R. and Dhillon, G. (2015). Dynamics of data breaches in online social networks: understanding threats to organizational information security reputation. *Proceedings of the Thirty Sixth International Conference on Information Systems*.

28 Karwatzki, S., Trenz, M., Tuunainen, V.K., and Veit, D. (2017). Adverse consequences of access to individuals' information: an analysis of perceptions and the scope of organisational influence. *European Journal of Information Systems* 26 (6): 688–715.

29 Acquisti, A., Friedman, A., and Telang, R. (2006). Is there a cost to privacy breaches? An event study. *Proceedings of the 27th International Conference on Information Systems*, p. 94,

30 Jang, I. and Yoo, H.S. (2009). Personal information classification for privacy negotiation. *2009 4th International Conference on Computer Sciences and Convergence Information Technology*, pp. 1117–1122. https://doi.org/10.1109/ICCIT.2009.322.

31 Yee, G. and Korba, L. (2005). Comparing and matching privacy policies using community consensus. *Proceedings of the 16th IRMA International Conference*, San Diego, CA, USA.

32 Courtney, K.L. (2008). Privacy and senior willingness to adopt smart home information technology in residential care facilities. *Methods of Information in Medicine* 47 (01): 76–81. https://doi.org/10.3414/ME9104.

33 Kumar, P., Braeken, A., Gurtov, A. et al. (2017). Anonymous secure framework in connected smart home environments. *IEEE Transactions on Information Forensics and Security* 12 (4): 968–979. https://doi.org/10.1109/TIFS.2016.2647225.

34 Lee, S., Kim, J., and Shon, T. (2016). User privacy-enhanced security architecture for home area network of Smartgrid. *Multimedia Tools and Applications* 75 (20): 12749–12764. https://doi.org/10.1007/s11042-016-3252-2.

35 Sivaraman, V., Gharakheili, H.H., Vishwanath, A. et al. (2015). Network-level security and privacy control for smart-home IoT devices. *2015 IEEE 11th International Conference on Wireless and Mobile Computing, Networking and Communications (WiMob)*, pp. 163–167. https://doi.org/10.1109/WiMOB.2015.7347956.

36 Cadwalladr, C. and Graham-Harrison, E. (2018). How Cambridge Analytica turned Facebook 'likes' into a lucrative political tool. *The Guardian*, March 2018. http://www.theguardian.com/technology/2018/mar/17/facebook-cambridge-analytica-kogan-data-algorithm (accessed 24 September 2021).

37 BBC (2017). German parents told to destroy Cayla dolls over hacking fears. *BBC News*, February 2017. https://www.bbc.com/news/world-europe-39002142 (accessed 24 September 2021).

38 Sarkar, S. (2017). Samsung Confirms its Smart TVs Listen & Transmit Everything You Speak, February 2017. http://techpp.com/2017/02/21/samsung-smart-tv-privacy-vulnerability/ (accessed 24 September 2021).

39 Proofpoint (2014). Proofpoint uncovers internet of things (IoT) cyberattack proofpoint US. https://www.proofpoint.com/us/proofpoint-uncovers-internet-things-iot-cyberattack (accessed 24 September 2021).

40 Lewis, J.A. (2016). *Managing Risk for the Internet of Things*. Center for Strategic & International Studies.

41 Razouk, W., Sgandurra, D., and Sakurai, K. (2017). A new security middleware architecture based on fog computing and cloud to support IoT constrained devices. *Proceedings of the 1st International Conference on Internet of Things and Machine Learning*, p. 35. ACM.

42 Samaila, M.G., Neto, M., Fernandes, D.A.B. et al. (2018). Challenges of securing Internet of Things devices: a survey. *Security and Privacy* 1 (2): e20.

43 Muhammad, R. (2019). Authentication and privacy challenges for internet of things smart home environment. *Journal of Mechanics of Continua and Mathematical Sciences* 14 (1). https://doi.org/10.26782/jmcms.2019.02.00018.

44 Stallings, W. (2017). *Network Security Essentials: Applications and Standards*, 6e, global edition. Pearson. ISBN 978-1-292-15485-5.

45 Mantas, G., Lymberopoulos, D., and Komninos, N. (2011). Security in smart home environment. In: *Wireless Technologies for Ambient Assisted Living and*

Healthcare: Systems and Applications (ed. A. Lazakidou, K. Siassiakos, and K. Ioannou), 170–191. IGI Global. ISBN 978-1-61520-805-0 978-1-61520-806-7.

46 Antonakakis, M., April, T., Bailey, M. et al. (2017). Understanding the Mirai Botnet. In: *26th USENIX Security Symposium (USENIX Security 17)*, 1093–1110. Vancouver, BC: USENIX Association. ISBN 978-1-931971-40-9. https://www.usenix.org/conference/usenixsecurity17/technical-sessions/presentation/antonakakis.

47 Kang, W.M., Moon, S.Y., and Park, J.H. (2017). An enhanced security framework for home appliances in smart home. *Human-centric Computing and Information Sciences* 7 (1): 6. https://doi.org/10.1186/s13673-017-0087-4.

48 Thalheim, B., Al-Fedaghi, S., and Al-Saqabi, K. (2008). Information stream based model for organizing security. *2008 3rd International Conference on Availability, Reliability and Security*, pp. 1405–1412. IEEE.

49 Kalaimannan, E. and Gupta, J.N.D. (2017). The security development lifecycle in the context of accreditation policies and standards. *IEEE Security & Privacy* 15 (1): 52–57.

50 Li, Y. (2013). Design of a key establishment protocol for smart home energy management system. *2013 5th International Conference on Computational Intelligence, Communication Systems and Networks*, pp. 88–93, Madrid, Spain, June 2013. IEEE. ISBN 978-1-4799-0587-4 978-0-7695-5042-8. https://doi.org/10.1109/CICSYN.2013.42. http://ieeexplore.ieee.org/document/6571348/.

51 Salami, S.A., Baek, J., Salah, K., and Damiani, E. (2016). Lightweight encryption for smart home. *2016 11th International Conference on Availability, Reliability and Security (ARES)*, pp. 382–388. Salzburg, Austria, August 2016. IEEE. ISBN 978-1-5090-0990-9. https://doi.org/10.1109/ARES.2016.40. http://ieeexplore.ieee.org/document/7784596/.

52 Kim, Y.-P., Yoo, S., and Yoo, C. (2015). DAoT: Dynamic and energy-aware authentication for smart home appliances in Internet of Things. *2015 IEEE International Conference on Consumer Electronics (ICCE)*, pp. 196–197. Las Vegas, NV, USA, January 2015. IEEE. ISBN 978-1-4799-7543-3. https://doi.org/10.1109/ICCE.2015.7066378. http://ieeexplore.ieee.org/document/7066378/.

53 Mantoro, T., Ayu, M.A., and binti Mahmod, S.M. (2014). Securing the authentication and message integrity for smart home using smart phone. *2014 International Conference on Multimedia Computing and Systems (ICMCS)*, pp. 985–989. https://doi.org/10.1109/ICMCS.2014.6911150.

54 Yu, Q., Zhang, C.N., and Huang, X. (2010). An RC4-based hash function for ultra-low power devices. *2010 2nd International Conference on Computer Engineering and Technology*, Volume 1, pp. V1–323–V1–328. https://doi.org/10.1109/ICCET.2010.5486127.

55 Chakravorty, A., Wlodarczyk, T., and Rong, C. (2013). Privacy preserving data analytics for smart homes. *2013 IEEE Security and Privacy Workshops*, pp. 23–27. San Francisco, CA, May 2013. IEEE. ISBN 978-1-4799-0458-7. https://doi.org/10.1109/SPW.2013.22. http://ieeexplore.ieee.org/document/6565224/.

56 Venda, P. (2015). Hacking Defcon 23's IoT village Samsung fridge — pen test partners. https://www.pentestpartners.com/security-blog/hacking-defcon-23s-iot-village-samsung-fridge/ (accessed 24 September 2021).

57 Torgilsman, C. and Bröndum, E. (2020). Ethical hacking of a Robot vacuum cleaner.

58 Transparent Proxying (2021). Mitmproxy - an interactive https proxy. https://mitmproxy.org/ (accessed 24 September 2021).

59 Chifor, B.-C., Bica, I., Patriciu, V.-V., and Pop, F. (2018). A security authorization scheme for smart home Internet of Things devices. *Future Generation Computer Systems* 86: 740–749. https://doi.org/10.1016/j.future.2017.05.048.

60 Notra, S., Siddiqi, M., Gharakheili, H.H. et al. (2014). An experimental study of security and privacy risks with emerging household appliances. *2014 IEEE Conference on Communications and Network Security*, pp. 79–84. San Francisco, CA, USA, October 2014. IEEE. ISBN 978-1-4799-5890-0. https://doi.org/10.1109/CNS.2014.6997469. http://ieeexplore.ieee.org/lpdocs/epic03/wrapper.htm?arnumber=6997469.

61 Jose, A.C., Malekian, R., and Ye, N. (2016). Improving home automation security; integrating device fingerprinting into smart home. *IEEE Access* 4: 5776–5787. https://doi.org/10.1109/ACCESS.2016.2606478.

62 Seigneur, J.-M., Jensen, C.D., Farrell, S. et al. (2003). Towards security auto-configuration for smart appliances. *Proceedings of the Smart Objects Conference*, Volume 2003.

63 Nobakht, M., Sivaraman, V., and Boreli, R. (2016). A host-based intrusion detection and mitigation framework for smart home IoT using OpenFlow. *2016 11th International Conference on Availability, Reliability and Security (ARES)*, pp. 147–156. Salzburg, Austria, August 2016. IEEE. ISBN 978-1-5090-0990-9. https://doi.org/10.1109/ARES.2016.64. http://ieeexplore.ieee.org/document/7784565/.

64 Monie, A. (2016). New Chromecast and Chromecast Audio. Have They Fixed Their Hijacking Issue? — Pen Test Partners. https://tinyurl.com/y8c77ddm (accessed 24 September 2021).

65 Whittaker, Z. (2019). Techcrunch is now a part of verizon media. https://techcrunch.com/2019/01/02/chromecast-bug-hackers-havoc/ (accessed 24 September 2021).

66 William, T. (2021). Hackers Breach Thousands of Security Cameras, Exposing Tesla, Jails, Hospitals. *Bloomberg.com*, March 2021. https://www.bloomberg .com/news/articles/2021-03-09/hackers-expose-tesla-jails-in-breach-of-150-000-security-cams (accessed 24 September 2021).

67 Filip, K. (2021). Verkada Security Update. https://www.verkada.com/security-update/ (accessed 24 September 2021).

68 Bradley, B. (2021). Privacy concerns raised after massive surveillance camera breach, March 2021. https://www.scmagazine.com/home/security-news/iot/ camera-tricks-privacy-concerns-raised-after-massive-surveillance-cam-breach/ (accessed 24 September 2021).

69 Bagüés, S.A., Zeidler, A., Valdivielso, C.F., and Matias, I.R. (2007). Sentry@ home-leveraging the smart home for privacy in pervasive computing. *International Journal of Smart Home* 1 (2): 129–145.

70 Cho, J.-S., Park, S.-H., Han, Y.-J., and Chung, T.-M. (2007). CAISMS: A context-aware integrated security management system for smart home. *The 9th International Conference on Advanced Communication Technology*, pp. 531–536. Gangwon-Do, Korea, February 2007. IEEE. ISBN 978-89-5519-131-8. https://doi.org/10.1109/ICACT.2007.358411. http://ieeexplore.ieee.org/ document/4195190/.

71 Zeng, E., Mare, S., and Roesner, F. (2017). End user security and privacy concerns with smart homes. *13th Symposium on Usable Privacy and Security (${\$SOUPS\$}\$ 2017)*, pp. 65–80.

72 Liberatore, M. and Levine, B.N. (2006). Inferring the source of encrypted http connections. *Proceedings of the 13th ACM Conference on Computer and Communications Security*, CCS '06, pp. 255–263. New York, NY, USA. Association for Computing Machinery. ISBN 1595935185. https://doi.org/ 10.1145/1180405.1180437.

73 OConnor, T.J., Enck, W., and Reaves, B. (2019). Blinded and confused: uncovering systemic flaws in device telemetry for smart-home internet of things. *Proceedings of the 12th Conference on Security and Privacy in Wireless and Mobile Networks*, WiSec '19, pp. 140–150. New York, NY, USA: Association for Computing Machinery. ISBN 9781450367264. https://doi.org/10.1145/3317549. 3319724.

74 Notra, S., Siddiqi, M., Gharakheili, H.H. et al. (2014). An experimental study of security and privacy risks with emerging household appliances. *2014 IEEE Conference on Communications and Network Security*, pp. 79–84. https://doi.org/10.1109/CNS.2014.6997469.

75 Fabian, B. and Feldhaus, T. (2014). Privacy-preserving data infrastructure for smart home appliances based on the octopus DHT. *Computers in Industry* 65 (8): 1147–1160. https://doi.org/10.1016/j.compind.2014.07.001.

76 Hoang, N.P. and Pishva, D. (2015). A TOR-based anonymous communication approach to secure smart home appliances. *2015 17th International Conference on Advanced Communication Technology (ICACT)*, pp. 517–525. Phoenix Park, PyeongChang, South Korea, July 2015. IEEE. ISBN 978-89-968650-5-6. https://doi.org/10.1109/ICACT.2015.7224918. http://ieeexplore.ieee.org/document/7224918/.

77 Dingledine, R., Mathewson, N., and Syverson, P. (2004). Tor: The second-generation onion router. *13th USENIX Security Symposium (USENIX Security 04)*, San Diego, CA, August 2004. USENIX Association. https://www.usenix.org/conference/13th-usenix-security-symposium/tor-second-generation-onion-router.

78 Apthorpe, N., Reisman, D., and Feamster, N. (2017). A Smart Home is No Castle: Privacy Vulnerabilities of Encrypted IoT Traffic. *arXiv:1705.06805 [cs]*. http://arxiv.org/abs/1705.06805. arXiv: 1705.06805.

79 Copos, B., Levitt, K., Bishop, M., and Rowe, J. (2016). Is anybody home? Inferring activity from smart home network traffic. *2016 IEEE Security and Privacy Workshops (SPW)*, pp. 245–251. San Jose, CA, May 2016. IEEE. ISBN 978-1-5090-3690-5. https://doi.org/10.1109/SPW.2016.48. http://ieeexplore.ieee.org/document/7527776/.

80 Burgess, M. (2018). Google Home's data leak proves the IoT is still deeply flawed. *Wired UK*, June 2018. ISSN 1357-0978. https://www.wired.co.uk/article/google-home-chromecast-location-security-data-privacy-leak (accessed 24 September 2021).

81 Craig, Y. (2018). Google's Newest Feature: Find My Home, June 2018. https://www.tripwire.com/state-of-security/vert/googles-newest-feature-find-my-home/ (accessed 24 September 2021).

82 Bowles, N. (2018). Thermostats, locks and lights: digital tools of domestic abuse. *The New York Times*, June 2018. ISSN 0362-4331. https://www.nytimes.com/2018/06/23/technology/smart-home-devices-domestic-abuse.html (accessed 24 September 2021).

83 vpnMentor (2019). Report: Orvibo Smart Home Devices Leak Billions of User Records. https://www.vpnmentor.com/blog/report-orvibo-leak/ (accessed 24 September 2021).

84 Tomáš, F. (2019). Two billion user logs leaked by smart home vendor, July 2019. https://www.welivesecurity.com/2019/07/02/two-billion-logs-leaked-smart-home/ (accessed 24 September 2021).

85 Palekar, M., Fernandes, E., and Roesner, F. (2019). Analysis of the susceptibility of smart home programming interfaces to end user error. *2019 IEEE Security and Privacy Workshops (SPW)*, pp. 138–143. San Francisco, CA, USA, May 2019. IEEE. ISBN 978-1-72813-508-3. https://doi.org/10.1109/SPW.2019.00034. https://ieeexplore.ieee.org/document/8844640/.

86 Nandi, C. and Ernst, M.D. (2016). Automatic trigger generation for rule-based smart homes. *Proceedings of the 2016 ACM Workshop on Programming Languages and Analysis for Security*, pp. 97–102. Vienna Austria, October 2016. ACM. ISBN 978-1-4503-4574-3. https://doi.org/10.1145/2993600.2993601. https://dl.acm.org/doi/10.1145/2993600.2993601.

87 Surbatovich, M., Aljuraidan, J., Bauer, L. et al. (2017). Some recipes can do more than spoil your appetite: analyzing the security and privacy risks of IFTTT recipes. *Proceedings of the 26th International Conference on World Wide Web*, pp. 1501–1510. Perth Australia, April 2017. International World Wide Web Conferences Steering Committee. ISBN 978-1-4503-4913-0. https://doi.org/10.1145/3038912.3052709.

88 Hron, M. (2018). At least 32,000 smart homes and businesses at risk of leaking data | Avast. https://blog.avast.com/mqtt-vulnerabilities-hacking-smart-homes (accessed 24 September 2021).

89 Scott, A. (2018). Security Researchers Find Vulnerable IoT Devices and MongoDB Databases Exposing Corporate Data, September 2018. http://blog.shodan .io/security-researchers-find-vulnerable-iot-devices-and-mongodb-databases-exposing-corporate-data/ (accessed 24 September 2021).

90 Younis, Y.A. and Kifayat, K. (2019). An infrastructure detection to cache side-channel attacks in cloud computing. *International Conference on Technical Sciences (ICST2019)*, Volume 6, p. 04.

91 Morrow, T., LaPiana, V., Faatz, D., and Hueca, A. (2019). *Cloud Security Best Practices Derived from Mission Thread Analysis*. Carnegie Mellon University.

10

IoT Hardware-Based Security: A Generalized Review of Threats and Countermeasures

Catherine Higgins, Lucas McDonald, Muhammad Ijaz Ul Haq, and Saqib Hakak

Department of Computer Science, Faculty of Computer Science, University of New Brunswick, Information Technology Centre, NB E3B 5A3, Canada

10.1   Introduction

Security in the Internet of Things (IoT) is paramount as the number of IoT devices are on the rise and new applications are emerging. We can find applications of IoT devices in the field of medicine, the automobile sector, personal devices, and more. Threats include breaking the security of devices to either steal or manipulate data or to destroy the devices altogether. Attacks on IoT devices can have serious consequences especially when they can compromise the safety of an individual. An attacker could theoretically alter medical data resulting in a patient being misdiagnosed or given the wrong medical treatment. There are many reasons why such devices are vulnerable and securing IoT devices is a challenge. IoT devices are attractive targets for collecting personal information as huge amounts of data are sensed, collected, consolidated, and analyzed by IoT devices [1]. Sensors are also easily accessible and often have little surveillance making it easy to destroy or tamper with [2]. Further, IoT devices are oftentimes interconnected together to form a network. Any compromise to one device in such a network can put the whole network at risk [2]. IoT devices are also easily exploitable due to their lack of computational and storage capabilities [2]. These are all some of the reasons that make securing the hardware as important as software.

Attacks on IoT devices are often classified by a four-layer architecture that includes perception layer, network layer, middleware layer, and application layer or by its application field [1]. This four-layer architecture is an abstraction of hierarchical layers wherein each layer holds unique responsibilities and is vulnerable to a variety of attacks [1]. The perception layer which is the focus

Security and Privacy in the Internet of Things: Architectures, Techniques, and Applications,
First Edition. Edited by Ali Ismail Awad and Jemal Abawajy.

of this chapter deals with the hardware components such as radio frequency identification (RFID) tags, cameras, sensors, and wireless sensor network (WSN) [1]. Researchers [2] claim that securing hardware devices is at the heart of IoT security. This suggests that the security of IoT can only truly be achieved if the underlying hardware devices are secure. However, few research studies consider other aspects of IoT critical [1].

We have divided the chapter into the following categories: Section 10.2 presents different types of hardware attacks. Section 10.3 highlights the countermeasures to mitigate such attacks. Section 10.4 concludes the chapter and Section 10.4 acknowledges the contributions of each author. The taxonomy of the attacks is presented in Figure 10.1. These threats are divided into three main categories, which are Hardware Design threats, Side Channel attacks, and Node Level threats.

10.2 Hardware Attacks

In this section, we have provided a brief discussion on different IoT security attacks.

10.2.1 IoT Devices

IoT devices are small-programmed pieces of hardware that can transmit and receive data to perform a task that are typically connected to the Internet and/or other devices. Two major technologies enable these devices to transmit or receive data which are RFID and WSNs [3].

WSN consists of several interconnected IoT devices also called nodes that work cooperatively to achieve a shared common goal [4]. These networks consist of nodes that contain sensors that collect information from the environment or from the control units that collect information from the sensors [5]. Sensors enable IoT devices to sense the environmental and physical state of a system [4]. On the other hand, the main purpose of RFID is to track and identify objects [4]. RFID technology is discussed in greater detail in Section 10.2.1.2.

The general architecture of IoT devices work on three layers. The first layer involves the perception layer which involves the sensors. This layer purpose is to acquire the data and transmit it to the network. The second layer is the network layer which transmits data over a network by employing technology such as WiFi, Bluetooth, and 3G. Lastly, the application layer is responsible for the authenticity, integrity, and confidentiality of data [6].

In terms of hardware-based attack, the IoT node is itself vulnerable to two categories of threats which are Node-Level and RFID-based [5]. In the following subsections, both these categories are explored.

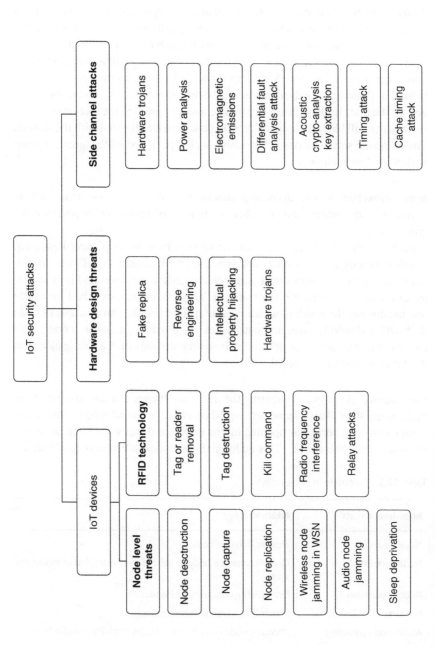

Figure 10.1 Taxonomy of IoT hardware attacks.

10.2.1.1 Node-Level Threats

As the focus of this chapter is on the hardware security of a device, this section will progress from the most simple threats within node hardware security and slowly progress to the more nuanced attacks. As such it will examine Node Destruction, Node Capture, Node Replication, Wireless Node Jamming, Audio Node Jamming, and Sleep Deprivation. Attack strategy for Node-Level based threats is also highlighted in Table 10.1. As the dangers listed below stem from physical access to the area or the node itself, within a defensive context examining the section on physical security will likely provide the most benefit once familiarized with the subject. The following subcategories will examine each type of node-level threat, starting with Node Destruction.

Node Destruction In the upcoming attacks, the security of the node itself is examined, and understandably, this is an area of particular importance for end-nodes [7].

As nodes may be in remote areas or regions hard to keep surveillance on, physical damage can be a costly and effective attack against an IoT network [8]. A physical attack requires direct access to the node in question and is as simple as an individual destroying the node [8]. This can be accomplished by nearly any means, but the result of which is an inoperable device. An example of node destruction should be clear and constitutes any physical assault on a node. This can also include more sophisticated destruction methods such as using electrical discharge to destroy a node [8].

Node Capture By physically altering the node, an attacker can take control of the node in question. This can be accomplished either by connecting to the device through an open port or by opening up the node and directly wiring to the board or components therein [8]. This additionally includes using tools to physically

Table 10.1 Summary of node attacks.

Node-level threats	Attack strategy
Node destruction	Destruction of a node
Node capture	Compromising a node to behave in unintended or malicious ways
Node replication	Introducing compromised nodes
Wireless node jamming	Preventing wireless communication
Audio node jamming	Prevents node from functioning by exploiting audio input
Sleep deprivation	Prevents node from entering sleep state

alter the memory of a node. A node captured or compromised could affect users in a multitude of ways [8], for example a compromised security camera could provide a feed that shows no one is present within an area when in reality the opposite is true (false data) [8]. Alternatively, a compromised node may launch a software-based attack on the network such as a denial-of-service attack [8]. A compromised node not only costs a company the asset, but it also could damage the integrity or availability of the entire system while interacting with other nodes (see Node replication) [8]. It should be stressed that as there could be some ambiguity, the difference between a Node Capture attack and a Hardware Trojan is as follows: Unlike a Hardware Trojan, a Node Capture Attack is done by compromising nodes rather than compromising the process in which the IoT devices are created. In other words, Node Capture describes a node that was compromised while deployed, while a Hardware Trojan describes a node that is compromised before it was deployed.

Node Replication   Before describing this attack, it should be explained that within an IoT network of end nodes, the nodes themselves tend to be homogeneous [7]. As such, the nodes tend to operate and work similarly.

Similar to node capture, node replication describes an attack that requires an IoT network being injected with multiple compromised nodes [8]. Node replication can have all the effects of a node capture attack as well, as more compromised nodes may allow greater control over the traffic of the IoT network, leading to the ability to suppress information or create false information on a grander scale [8].

Wireless Node Jamming in WSN   A WSN is a network consisting only of sensor nodes [9]. The goal of a wireless jamming attack is to prevent communication from different nodes by ensuring that wireless communication is impossible or extremely slow [10]. As such, this is an attack on the availability of a device. This can be done by simply producing noise on the used channel [10]. Li et al. [10] described what to look for to determine if a system is at risk of a Wireless Node Jamming Attack:

(1) Systems in which an adversary does not need special hardware to launch the jamming attack.
(2) Systems where the attack can be implemented by simply listening to the open medium and broadcasting in the same frequency band as the network uses.
(3) Systems where there are significant benefits for the attacker while maintaining little cost to the attacker.

When implementing nodes, networks that use wireless networks to communicate either between nodes or to a central hub, it would be pertinent to keep these considerations in mind especially when evaluating the risk posed to the network.

Audio Node Jamming A precursor attack to Audio Node Jamming attack was first described by Zhang et al. [11] and coined as a dolphin attack. A dolphin attack makes use of inaudible sounds to trigger audio inputs on a device and remotely run audio commands. Elaborating on the aforementioned Dolphin Attack, the work of [12] showed that by using similar techniques they were able to prevent audio inputs from being identified by simply producing random noise beyond the hearing capabilities of the human ear. As the Dolphin attack refers to a specific attack that takes advantage of both hardware and software, in this case, the attack described by Mao et al. [12] refers simply to preventing audio input from being read by a node.

Though novel, this attack has been included as an example of an alternate form of jamming that may occur against individual nodes. Although wireless network-based jamming attacks or RFID attacks may seem like an obvious target for attacks, other input mediums could potentially be attacked as well and so caution should be taken when addressing the security of IoT end nodes.

Sleep Deprivation IoT nodes can be placed in areas where it is hard to access them and because of this they cannot be easily serviced [13]. As it is difficult to service the node, the resources of a node may become incredibly important to maintain for the IoT node network to function properly [13]. For example, a device may use a wireless transmitter to relay important data concerning the current temperature in a forest, when it reaches a certain level, it may send a signal that there may be a forest fire. A denial of sleep attack (DoSA) is particularly dangerous against devices such as these which are difficult to service [13]. The attack occurs when a malicious attacker wakes sleeping devices repeatedly; the implementation of this varies based on the protocol of the device used. In Media Access Control (MAC) protocol-based wireless systems, this attack occurs by introducing a new node and having it periodically transmit to other legitimate nodes, preventing them from entering a sleep mode [13]. In the forest example, this IoT temperature sensor likely only has the bare minimal specifications needed to accomplish its goal as such battery life-impacting attacks are of significant concern [14]. If successful, the temperature node may simply run out of batteries, preventing it from alerting the authorities when a fire breaks out.

10.2.1.2 RFID Technology

RFID is an integral technology in IoT [15] and is growing in popularity as it plays an important role in many applications by identifying and tracking objects of interest. Some common examples of applications are in the areas of navigation, healthcare, military, and asset management [16]. For example in the healthcare sector, RFID technology allows for the monitoring of patient's health and real-time disease management [17].

RFID technology consists of two parts: RFID tags and RFID readers/interrogators [16]. RFID tags consist of an antenna, one or more integrated circuits (ICs) and an Electronic Product Code (EPC) for identification purposes [15, 18]. There are different types of tags: active, semi-passive, and passive [16]. Generally, passive tags are cheaper and as a result, the more popular as they are more energy-efficient. On the other hand, active RFID is more expensive but have their own power source and a wider range of functionality [16].

In RFID technology, there are two avenues of communication. First, the readers communicate with the server to store information either in the cloud or a database system [17]. Second, the reader and the tag communicate with each other; the RFID readers can pass information by means of a radio-frequency signal and in turn, the readers can obtain signals from the tags [16]. There are various threats to both these communication channels as either is not secure. Further, RFID technology is susceptible to a wide range of attacks and is open to threats because of their unstable or noisy environment that they are often found in [18]. Moreover, RFID technology is developing rapidly creating an evolving threat landscape as well [18]. In addition, there is no physical security in place to protect RFID technology from attacks [19].

Threats to Tags Permanently disabling tags is one threat to tags which can happen by tag removal, tag destruction, or using a KILL command [18]. Some motivations behind disabling tags are to make them untraceable and unusable. Any IoT device could then easily be stolen. Anyone can remove a tag if it is not embedded as it does not require a great amount of skill to perform [19]. Destroying tags is also easy. An attacker can apply pressure, clip the antenna, or use chemicals to destroy the tag [19]. Tags are also sensitive to their environment. For example extreme cold or hot temperatures can destroy a tag [19]. In addition, discharging the battery of active RFID tags can render them obsolete [19]. Electrostatic discharge or high-energy waves can disable RFID tags as well [19].

The KILL command is another way to disable RFID tags. The design of this command is to permanently silence an RFID tag [19]. Tags become unusable by using the password given to each tag by the manufacturer [19]. Database information can also be completely or partially deleted with the execution of the KILL command [19].

Threats to Readers Like RFID tags, attackers may remove RFID readers as well, especially if left unsupervised [19]. Tags and readers ideally require mutual authentication as an unauthorized reader can read sensitive information from a tag [15]. In order to gain information about tags, an attacker may target an RFID reader for cryptographic information. Further, the attacker could manipulate information on the back-end [19].

Radio Frequency Interference Tags and readers communicate with each other by means of radio-wave signals. The passive RFID tag must rely on the reader for its own reflection power since a passive RFID tag has no internal power source [16]. A reader activates an RFID tag by its forward/transmission power and in turn, the tag reflects the power back to the reader [16]. The power which the tag receives from the reader is more powerful than what the tag reflects back to the reader [16]. The tag then must reflect enough power back to the reader [16]. In a passive network, the reflection power of all the RFID tags creates an overall interference which then drives the tags to compete with each other to reflect more power [16]. An attacker can easily introduce interference by adding an intruder tag into the network with high interference, leaving other tags hard for readers to pick up [16]. This is also referred to as active jamming and is extremely disruptive to the whole network [19]. Other natural inference from sources such as power switching supplies or electronic generators can also happen that interfere with the communication [19].

Relay Attacks A relay attack is a man-in-the-middle type of attack where an attacker places a device between the reader and the tag [19]. The attacking device that is between the tag and the reader is able to trick the reader and the tag to think that they are communicating directly with one another by intercepting and modifying the radio signal [19]. In this attack, the attacker is only using one device; however, an attacker can use two devices, one to communicate with the tag and one to communicate with the reader [19]. Also, for this attack, distance is not a factor once a relay has been intercepted, as an attacker can use the relayed data at their leisure at any point afterward [19]. For example, after having gained the victims RFID information, an attacker may use that information to purchase products in a completely different country.

Table 10.2 presents an overview of RFID-technology-based threats and their attack strategies.

Table 10.2 Summary of RFID threats.

RFID attack name	Attack strategy
Tag or reader removal	Removal of tag or reader requires no specialized knowledge
Tag destruction	Destruction of tag similar to removal can be easy to accomplish
Kill command	Using the password created by the manufacturer to permanently disable an RFID tag
Radio frequency interference	Preventing radio communication between tags and readers
Relay attacks	Man-in-the-middle attack: a device intercepts information and relays information back to a reader

10.2.2 Hardware Design Threats

An IC is a small microchip that allows smart devices to perform communication and data processing [2]. Outsourcing the production of this microchip is common as a result of the increased demand for IC boards and globalization [2]. This leaves hardware open to different types of vulnerabilities such as fake replicas, reverse engineering (RE), Intellectual Property (IP) hijacking, and Hardware Trojans (HTs) [20].

Hardware design attacks can be further categorized into the following subtypes (see Table 10.3 for a summary):

10.2.2.1 Fake Replica

Fake Replica is a type of attack that concerns the IP right owners. It involves the counterfeit of the IP [20]. Counterfeits are illegal and a major issue in the microelectronics industry [21]. There is a multitude of ways to counterfeit such as relabeling, refurbishing, and repackaging. Relabeling involves replacing identifiers to mislead buyers into thinking a product is from a known supplier or of a higher grade [21]. An attacker can reuse old or expired designs and sell them as a new process known as refurbishing [21]. Repackaging is changing the current packaging of an IC [21]. Counterfeiting can have harmful consequences. Often times in the process of counterfeiting, there is damage to the product such as alterations to the circuit due to poor handling conditions [21]. The handling of the IC may result in a nonfunctioning chip [21]. Many times, counterfeiting can be used for financial gain, or for malicious intent, as a bad-faith counter fit may introduce a malicious circuit known as a HT [20]. HT in turn compromises the functioning of the IC and can have serious consequences on devices that use them, such as vehicles and airplanes [20].

10.2.2.2 Reverse Engineering

RE is the process of deconstructing a device in order to gain an understanding of how the device works. An attacker may reverse engineer an IC chip in order

Table 10.3 Summary of hardware design threats.

Hardware design threats	Attack strategy
Fake replica	Relabeling, repackaging, or using old or expired designs
Reverse engineering	Deconstructing device
Intellectual property hijacking	Bypassing copyright and stealing design
Hardware Trojans	Change expected function or add new malicious functions

to later reconstruct it and add a malicious circuit [20]. RE also enables attackers to find weaknesses in the design for attack purposes [21]. Other reasons for RE include stealing the design or revealing Intellectual Property information about the design.

10.2.2.3 Intellectual Property Hijacking

Intellectual property hijacking involves the theft of design information without respecting the copyrights [20]. An example of this is where an attacker who manufactures a copyright protected chip overbuilds the chip in order to sell it through the black market [20]. The consequences of design theft can include building the IC with extra circuits [21]. The attackers can also claim to be the owner of the new ICs built from the stolen design [21].

10.2.2.4 Hardware Trojans

HTs consist of malicious insertion or inclusion to an IC [22]. The intent of HT is to change the current normal functioning of the chip or to create a new malicious function [22]. HTs are hard to detect as they are intentionally designed this way and remain dormant until activated when a condition is met. HT are difficult to remove, unlike software Trojans that require a firmware update [2, 22]. HT have serious consequences in the fields of healthcare, military, and finances [23]. For example, a malfunction in life-saving equipment, a loss of control of missiles, or a leakage of cryptography keys are all serious consequences of malicious insertion [23]. Insertion of HT can happen at any point of the design stage [2]. Due to the increased demand in IC boards, companies are frequently outsourcing the production to cheaper factories leaving the chips even more vulnerable to HT insertions [2].

Generally, the activation of a HT happens when a condition set by the attacker is met [23]. Setting the trigger can happen internally or externally [23]. Conditions can include a given input pattern, an internal logic state or a counter value [23].

10.2.3 Side-Channel Attacks

In side-channel attacks, the devices reveal sensitive information unintentionally in a number of ways. The physical state of electromagnetic (EM) emission, power consumption, sound, and timing values are all examples of how devices leak information, side-channel attacks exploit these vulnerabilities. Attackers retrieve this data by running tests during normal operation [20]. Side-channel attacks are especially common in attacks on cryptographic implementations, as a means to either break or bypass the device's protection layers [20]. For example, an encryption algorithm produces a ciphertext when provided with a plain text and encryption key. The ciphertext is the intentional specific output; however, there is an unintentional output that is the by-product of the algorithm computations [24]. Variations in power consumption or execution time during the computation produce unintentional information leakage [24]. Analysis of this information allows attackers to detect patterns and consequently reveal private information.

In the past, in order to observe and learn information from the devices, the attacker needed physical possession of the device [24]. However, that is no longer the case as attacks can be performed remotely [24]. Side-channel attacks have seen a rapid development in both scope and scale since the early 2000s [24]. To classify the differences in types of attacks and in levels of invasiveness, there are two axes of classifications for side-channel attacks [24].

Axis one: active vs. passive. The first axis classifies attacks as either active or passive. An active attack is when the attacker modifies the behavior of the device by exploiting the inputs of the side-channel [2]. A side-channel refers to the physical or logical properties of a device [24]. The introduction of an error in the computation by means of fault-injection is one example of an active attack [2].

In contrast, a passive attack is when an intruder simply observes and learns from the device passively. This is to establish patterns and involves the output of the side-channel [2]. A person from a distance can capture information about the produced EM radiation and then analyze this data to reveal information presented on a screen [2]. This type of attack is called an electromagnetic analysis (EMA) attack.

Axis two: invasive vs. semi-invasive vs. noninvasive. The second axis classifies the level of invasiveness of a side-channel attack. This simply refers to the level of manipulation on a device [24].

10.2.3.1 Types of Side-Channel Attacks
In this subsection, we have briefly highlighted different types of side-channel attacks and their attack strategy as shown in Table 10.4.

Hardware Trojan Side-Channel Attacks HT side-channels are a type of HT that results in leakage of information. Modifying a few gates of an IC chip is a typical insertion of an HT [22]. Ender et al. [22] describes an HT that increases path delay by modifying a few gates. Detection of this HT is difficult because no logic is added to or removed from the gates. In ASIC platforms, subtle modification of the transistors at the subtransistor level can introduce an HT [22]. In Field-Programmable Gate Array (FPGA) platforms, a change in the route of certain signals can add an HT [22]. Once the device clock frequency increases beyond a certain

Table 10.4 Summary of side-channel threats.

Side-channel attacks	Attack strategy
Hardware Trojans	In this context modifies hardware gates to gain timing information
Electromagnetic emissions	Measuring electromagnetic waves to deduce information
Differential fault analysis/attack	Using fault ciphertexts and normal cipher texts to deduce information
Acoustic crypto-analysis key extraction attack	Measuring sound emitted from device to deduce information
Timing attack	Measuring time of operation to deduce information
Cache timing attack	Measuring time between cache and memory to deduce information

point, this triggers the activation of the HT and leaves the device vulnerable to side-channel attacks [22].

Power Analysis/Consumption Power analysis attack is a type of side-channel attack designed to gain access typically to cryptographic information by measuring power consumption. There are two groups of classification for this type of attack: simple power analysis (SPA) and differential power analysis (DPA) [2]. SPA attacks involve visual interpretation of power traces [25]. Traces are simple waveforms that are measurements of power consumption during cryptographic operations [25]. Traces provide information on the sequence of instruction execution and as a result, reveal cryptographic information [25]. For example, an SPA can determine when there is a jump instruction during code execution [25]. A DPA measures the differences in power consumption by examining changes to binary bits [25]. A DPA usually requires several samples to reveal information [25].

Electromagnetic Emissions Most electrical devices emit EM waves when in operation. By measuring these deductions, one can get information about the current state of the device and how it is operating [26]. As such it is important to examine attacks that stem from this EM information. When utilized this attack is a form of passive attack in that it requires no direct connection to the device itself and merely "listens" to the EM waves in order to learn more about the device itself. Unlike many attacks, EM emission attacks are being considered for use in the data recovery industry. Of specific interest to the data recovery industry is the growing relevance of data in systems that cannot be directly tampered with, such as a pacemaker [26]. However, as both adversary and legitimate enterprise may use these techniques, the scope of the vulnerability becomes clear. As such it is important to take away that although potentially valuable for legitimate enterprise, an attacker can use these wave emissions to determine the state of the device and infer information that could be used to attack the device and gain access to sensitive information [26].

Acoustic Crypto-Analysis Key Extraction Attack In 2017, Genkin et al. [27] published a paper detailing a new form of analysis attack. They showed that by using the actual sound emitted from a device when in operation, cryptographic information could be deciphered from the device. In particular, they showed that information such as different Rivest-Shamir-Adleman (RSA) keys produced different sounds and that within an hour they could extract the full 4096 bits of a key from a computer [27]. This attack is a great example of the creativity that may be employed by attackers. For this attack to work, the placement and sensitivity of the microphone is crucial, as this attack could be foiled by numerous outside factors including the noise of the room where the attack is taking place [27]. Within an IoT context, attacks like these could present themselves by an attacker placing a microphone or other device near an IoT device [27], this microphone could be hidden as it slowly gathers audio data during day-to-day operations, and feasible retrieved or abandoned once no longer needed. There is even the possibility that a mobile phone could be used to conduct this attack as theorized by Genkin et al. [27]. Acoustic Crypto-analysis may not become a prominent attack but as previously stated should be used as an example of the dangers that side channel attacks pose, as the simple emission of audio data in regular use could be used as a means to subvert security expectations.

Timing Attack A simple piece of information collected repeatedly and extrapolated can lead an attacker to gain access to a cryptographic system. First described by Kocher [28] and Lo'ai and Somani [29], the work of [28] describes the basis for understanding a timing attack. An attack that uses the time it takes to process a request to extrapolate hidden information. As per Kocher, the attack consisted of monitoring a channel used to transmit cryptographic information. Once the channel is monitored, the attacker takes precise measurements of the time it takes to respond. After which depending on the time used for each request, the attacker can extrapolate hidden information [28]. This sort of timing attack is sometimes referred to as a classical timing attack and is of particular relevance when discussing IoT [29]. It is important to note that for a classical timing attack to work, the information the user is attempting to extrapolate needs to be static between timings, as it will not work if the key dynamically changes, additionally, the attacker is expected to have the ability to gain information or already have information about the cryptographic system in place [28].

Differential Fault Analysis/Attack Differential Fault Analysis (DFA) poses a severe risk to IoT devices [30]. As a form of timing attack, a DFA attack focuses on the principle that by analyzing both a ciphertext created from a fault and a normal ciphertext, an attacker can deduce information about the private key [31]. In other words, by forcing a device to create a ciphertext while inducing faults in the system and comparing it to a ciphertext created in regular operation, they can begin to break down the security of the device itself. How an actual fault is introduced into a system varies, for example in some modern cases, two high-powered lasers are used on the board [32]. The scope of vulnerability for this attack appears to be quite large, as many cryptographic systems have shown at least some vulnerability to it [33]. Because of this, it is nearly impossible not to mention the effect of DFA within the context of IoT devices, as this attack can be carried out without the use of high-end equipment [30]. Adding to the risk is the fact that DFA attacks are relatively fast to carry out [31]. Given that the attack is effective, quick, and cost-effective, there is a real need to consider DFA as an attack that is going to define IoT security.

Cache Timing Attack A cache timing attack differs from a classical timing attack as it focuses on the timing difference between the CPU cache and the onboard memory [34]. The CPU cache is quite quick especially in comparison to onboard memory, as such both have different speeds [34]. Using this timing difference and the fact that different values may take different amounts of time, the attacker can extrapolate information about a system and potentially reveal secure information such as an encryption key [34].

10.3 Physical Security Attacks Countermeasures

Determining which IoT component deserves the most attention to ensure security of the device can be depicted by a pyramid of pain as seen in Figure 10.2 [35]. This pyramid of pain consists of layers that make up an IoT device and are stacked based on the level of impact if one of those layers were to be compromised. For example, sensors are the most accessible part of an IoT device and are vulnerable as a result; however, they have the least impact and are found at the top of the pyramid. On the other hand, the hardware layer is the least accessible layer but creates the greatest pain if compromised [2]. As the foundation of the pyramid is what supports the pyramid itself, the hardware layer is the most important factor in the security of IoT devices.

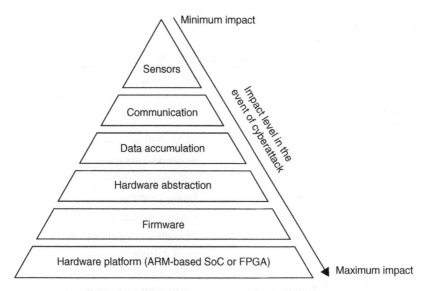

Figure 10.2 The Pyramid of Pain, illustrating the layers which have the greatest impact [35].

10.3.1 Mitigation Techniques for IoT Hardware Attacks

This section describes some of the major preventive measures for various hardware attacks on IoT devices. Table 10.5 provides an overall summary of each type of attack and the different strategies that can be employed.

10.3.2 Side-Channel Attacks

10.3.2.1 Hardware Trojans

One of the greatest threats to IoT device security is HTs. HTs are malicious logic or design flaws that are inserted into the ICs during the design phase [2]. These are difficult to find and remain dormant until they are triggered. HTs are not easy to remove as the only way to circumvent the Trojan is to either block the malfunctioning component or replace the device as a whole. As a result, securing IoT devices against a HT is a daunting task.

To guard against these kinds of attacks, an understanding of HTs is key. HT are best described based on several characteristics notably: their physical characteristics, insertion time/phase, the task they perform, or how they are triggered [2]. Figure 10.3 shows how a hardware Trojan with a trigger and payload can be added to a simple combinational circuit. When triggered, payload can change the correct functionality of the circuit.

Table 10.5 Types of IoT physical security threats.

IoT physical security threats		
Categories	**Threats**	**Preventive measures**
Side-channel attacks	Hardware Trojans	Obscuring design information
		Multimode operation
		Removing extra space
		Segregated production
	Power analysis	Hiding(confusion)
		Dummy operations
		Randomizing input data execution
		Clock delays
		Masking
	Timing attacks	Constant time operations
		Bucketing
		Random delays
		Refetching T-tables in Advanced Encryption Standard (AES)
	Electromagnetic analysis attacks	EM attack sensor
		High-frequency voltage regulator
	Acoustic analysis attack	Minimizing sound
		Generating parallel instructions
		Generating parallel sound
RFID attacks	Abuse of tag	Physical unclonable function
		Enhancing physical security
		Preventing information leakage
	Relay attacks	Distance bounding protocol
		Two-factor authentication
		Context-aware selective unlocking
Integrated circuit attacks	Counterfeit ICs	Physical unclonable functions
	Reverse engineering	

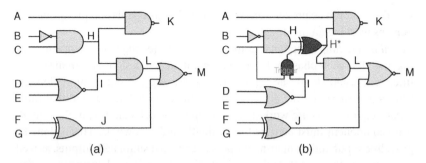

Figure 10.3 (a) A simple combinational circuit. (b) Trojan with trigger and payload added to the original circuit [36].

Figure 10.4 The different stages of the creation of an integrated circuit.

The hardware of an IoT device goes through various distinct stages. The first stage consists of the blueprint and IP. The second stage consists of the design and the last, the implementation as shown in Figure 10.4. In some cases, third parties do manufacturing, design, or the procurement of components. The need for a third party creates the possibility for a HT to be inserted into the IC at any stage. Any malicious actor with access to any phase of manufacturing can add a backdoor/HT into the IoT device, to be exploited at a later time [2]. Since the insertion of a HT can take place in any phases, designing countermeasures that prevent unauthorized manipulation of the hardware is key. An important countermeasure is detection.

Detection Mechanisms

a. *Design validation*. Design validation is a procedure that can be employed to ensure the design of the IoT device is not altered during its life cycle [2]. The design includes a schematic diagram, source code, Print Circuit Board (PCB) design and layout, and various components involved in the smooth running of IoT hardware. With design validation, it is crucial that after every phase of production that the device is compared with its original blueprint. This ensures that the layout of components, fabrication of the device, and logic gates used in the device are according to the original design [2]. If there are any changes during the validation, then it could be possible that there has been an HT insertion. Also, the original design itself must be thoroughly examined before it is sent to production facilities. Access to this blueprint must be kept to a minimum, and there should be proper access control and non-repudiation procedures for persons working on the design.

b. *Functional testing.* Functional testing includes the testing of IoT hardware during and after each phase of IoT development. When designs are finalized, the intended design should be rigorously tested using computer simulation [2]. This is to ensure that the device produces the expected output and that functional requirements are working according to the original design. Each hardware component must be tested individually and as a whole. This is done by providing input and comparing outputs to expected simulated outputs, as well as comparing the outputs to the original calculations derived during the design phase [2]. If there is more than one production facility involved in the process, then devices from each of these can be compared to each other, as well as the simulations to make sure that devices are the same as the original and have not been tampered with during production phase [2].

c. *Behavioral testing.* HT can avoid detection when conventional inputs and procedures are used to test the IoT hardware. It is possible to reason that the attacker may have the product test designs and can insert an HT in such a way to avoid detection during normal testing [2]. An effective means to test for HT is to test the behavior of IoT devices under unusual circumstances such as changing input sets, testing devices with variable input currents, testing devices under different operating temperatures, and finally observing the device behavior using previously known attacks on IoT devices [2]. These measures help designers and engineers to test the overall security of the device and trigger any hidden HT.

Prevention Mechanism The goal of prevention is to make the insertion of HT difficult for an attacker. A brief description of a few preventive approaches is given below:

a. *Obscuring design information.* Obscuring Design Information can be achieved by hiding the functional design of the IoT device as much as possible by adding nonfunctional or dummy components, jumpers, and contacts [2]. This makes it harder for attackers to insert HT into the IC. For example adding nonfunctional components to the IC increases the chances that an attacker inserts an HT into a nonfunctional component instead of a functional one rendering the HT obsolete [2].

b. *Multi-mode operation.* Multi-mode operation means that hardware is designed to work in multiple modes [2]. The designer adds an obfuscated mode that makes a device act in an unintelligent way in the hope of making it harder for an attacker to add an HT [2]. The two modes generate different outputs, making it difficult for an attacker to get the correct functionality of the device [2].

A device can only change mode when it receives an input only known by the designers and stay in obfuscated mode while not processing data.

c. *Removing extra space.* One way to make the insertion of HT difficult for an attacker is by eliminating any extra space in the IC [2]. Making the chip compact takes away space for an HT to be inserted in the first place. Circuits must be designed in such a way that any extra spaces are filled with active components [2].

d. *Segregated production.* Isolating each production module of the IoT device can help in preventing the insertion of HT. These modules can be given to one or more third parties to manufacture such that no third party has the complete design. Also, functional designs should be shared on a need-to-know basis only. This prevents an attacker from having the full knowledge of the design that might be needed for an HT. Ideally, any process where an HT insertion can occur should only be given to trusted parties. Sidhu et al. [2] mentions the process of split manufacturing in which the design is divided into Front End of Line (FEOL) and Back End of Line (BEOL) parts. The FEOL elements like transistors and capacitors are fabricated by an untrusted party, whereas BEOL parts consisting of higher-level interconnects are fabricated by a trusted party [2].

10.3.2.2 Power Analysis Attack

Different computations on a microprocessor consume different amounts of power. Power analysis attacks involve analyzing those changes in power consumption to reveal sensitive information. For example, power analysis attacks can be used to reveal a secret key by monitoring the power consumption of the targets cryptographic hardware. Figure 10.5 shows how a DPA attack can be used to deduce

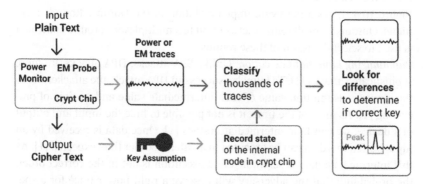

Figure 10.5 DPA (Differential power analysis) attack.

the secret key of the cryptographic circuit from monitored power traces. Security against these kinds of attacks is of critical importance. Countermeasures to thwart power analysis attacks can be at the gate level or algorithmic level [38]. Countering power analysis attacks involves hiding electrical consumption attributes as much as possible.

a. *Hiding (confusion).* In a power analysis attack, the attacker tries to map the type of data processed to the power consumed by the processing hardware. By altering the correct relationship between data and power consumption, we can make it difficult for an attacker to correctly guess the information being processed. This can be achieved by designing hardware in which every same operation consumes a random amount of power or all operations require the same amount of power [39].

b. *Dummy operations/rounds.* One type of power analysis attacks is a DPA attack. DPA attacks rely on input data, the specific time taken to process this input and the power consumption of the device at different stages of its processing input data. To launch a successful DPA attack, an adversary needs numerous power traces [39]. As IoT devices collect a huge amount of data, this gives attackers a large amount of power traces to infer the key or intermediate data. By adding random dummy operations in the process, the electrical consumption data will change. These dummy operations can be added anywhere in the process. Adding dummy operations can also help in achieving the goal of fixed power consumption for different input data, but it comes at the cost of performance. One of the countermeasures used against DPA attacks is round-based encrypting schemes which use dummy rounds [40]. Hardware is designed in such a way that multiple parts of hardware generate outputs, but their outputs are randomly used or not used for computation in every single clock cycle. This keeps the computations and power consumption the same for each clock cycle [40].

Figure 10.6 shows a hardware implementation of the Dummy Rounds technique. Outputs from all rounds act as input to a multiplexer. A round controller is used to select the result of these rounds.

c. *Randomizing input data execution.* Unlike SPA attacks, DPA attacks are more sophisticated attacks [1]. To launch a successful DPA attack, the attacker needs the power consumption value of each intermediate value and its time of processing [1]. As most of the time, it is not possible to hide the input and output of a device, we can hide intermediate values [1]. Once data is received by an input buffer, this data can be executed in a different order than expected [1]. As this intermediate data is scrambled and execution is not in the correct order, the probability that the adversary will observe a right power trace for a specific instruction is $1/N$, where N is the number of input instructions [1]. To get

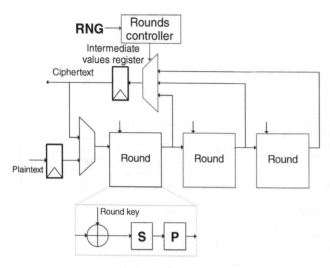

Figure 10.6 Dummy cycles countermeasure scheme [40].

the correct sequence of each instruction for an output, the original sequence is stored in a First In, First Out (FIFO) data structure to ensure that the output is in the same sequence as the input [1]. In this way, the execution of input data can be changed to hide processing information from the adversary. The larger the input size, the more difficult it is for the adversary to obtain correct power analysis of intermediate instructions [1].

Figure 10.7 shows the implementation of a random input data execution module. It consists of an input buffer, output buffer, Crypto core, and random address generator. Random address generator provides the functionality of scrambling the input sequence for randomized execution and preserving the right input sequence with the help of FIFO to restore the original sequence while outputting data.

d. *Clock delays*. In digital circuits, operations are dependent on the clock which dictates the correct sequence and time to execute instructions. By adding random delays in each cycle or adding one or more dummy cycles, it will make it harder for the adversary to analyze the correct power trace for a given input [41]. This can be implemented by using more than one phase-shifted clock in the circuit and randomly selecting a single clock input [42].

e. *Masking*. For secure communication and authentication, IoT devices have specialized cryptographic circuits implementing both Advanced Encryption Standard (AES) and Data Encryption Standard (DES) encryption schemes. However, these circuits are often targeted. An adversary using a power analysis attack can get a hold of the encryption keys. Masking is a method of adding a

Input data

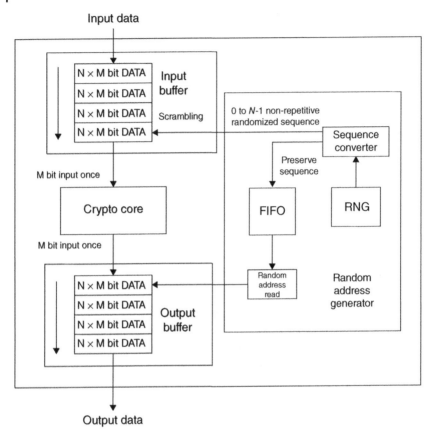

Figure 10.7 Illustration of the data processing module against DPA. Source: Adapted from [1].

random value to the input of these encryption schemes. For masking to work, these encryption schemes have to be redesigned to work with masked inputs [38]. Figure 10.8 shows a block cipher algorithm with masked inputs in each round. This can be achieved at the algorithmic-level and gate-level [38]. Algorithmic-level implementation involves rewriting encryption schemes to work with masked intermediate values while keeping input and output values the same as the original implementation [38]. Gate-level implementation of masking does not depend on the original implementation of the encryption scheme; it can be done by converting existing circuitry to use masked logic gates [38]. Gate-level implementation however introduces a new problem: digital circuit glitches. Glitches are unwanted gate transitions that occur due to a delay in intended gate input [43]. In one implementation of masking

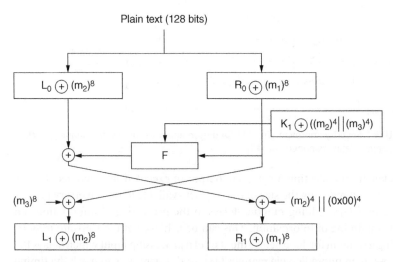

Figure 10.8 Masking method of defending differential power analysis attack in block cipher.

proposed by Mahanta et al. [44], a Masked RSA Scheme was designed that generated random numbers to mask the input values. Random values were also used to mask intermediate inputs at different stages to make power analysis difficult for the adversary [44].

10.3.2.3 Timing Attacks

Timing Attacks are a type of noninvasive attack on IoT devices that is attributed to increased attacks on these devices [45]. This type of attack can be carried out remotely [46]. IoT devices process input data and various intermediate data to produce the final output. As all inputs to the device take different time to process, and this time is dependent on the information the device is processing, an adversary can compare these time measurements and input data processed, with a simulated model to correctly guess intermediate data and execution order of the data inside the IoT device. Thus, for a successful timing attack, attackers try to find a relation between the input and execution time. Making time execution less dependent on the input can help to secure IoT devices against timing attacks.

a. *Constant time operations.* By making all inputs process in constant time, there is no information for the attacker to correctly guess system behavior. Jayasinghe et al. [47] proposed a constant round execution time for AES by rescheduling inputs, to counter against timing attacks on the AES encryption scheme.

b. *Bucketing.* Bucketing is a technique in which the outputs of executed instructions are stored in buckets. Buckets can be variable or fixed size. Buckets are only returned once they have reached their full capacity. This prevents an

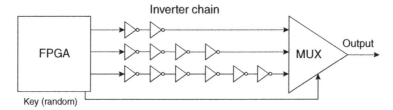

Figure 10.9 Security improvement of FPGA design against timing.side-channel attack using dynamic delay management [49].

attacker from collecting a complete output of executed instructions at once, this helps in reducing the size of the data set available to the adversary [45].

c. *Random delays.* Adding random delays in the processing of information can alter the timing of information. This can be achieved at the software level by adding a random for-loop before the start of a processing input which delays the process in an unpredictable manner [48]. In this way, we can mask the timing information. Figure 10.9 presents a hardware implementation for adding random delays which involve adding a variable number of NOR gates between a FPGA and a multiplexer [49]. By selecting randomly one of the paths, a random delay can be added to the process [49].

d. *Prefetching T-tables in AES implementation.* In T-table implementations, prefetching some of the T-table values into the cache memory before the start of the encryption process can change the pattern of cache accesses [48]. It is possible that these values can be removed from the cache if the memory is needed by other operations [48]. This change in cache access rate can change the pattern of encryption from the unaltered encryption process, as some of the values accessed from the cache can be missed by adversaries, making it difficult for adversaries to correlate timing and data processed [48]. Another way to increase cache accesses is by partitioning the cache and allocating specific locations for T-tables so that there is less chance that these values are not overwritten [48].

10.3.2.4 Electromagnetic Analysis Attacks

EMA attacks are a type of side-channel attacks. These attacks work like power analysis attacks by measuring the EM emission of the device and finding a relation between these EM emissions and processing data.

a. *Electromagnetic attack sensor.* EM readings are recorded using an EM probe to measure the EM field data. One of the solutions proposed by Miura et al. [50] is a hardware-based approach to embed an EM probe sensor in the circuit to detect when a probe is brought near the circuit of an IoT device. A fundamental principle of electromagnetism is that we cannot measure EM-fields without disturbing them. As such by having an EM emitter and sensor on the IoT device's circuit can help us to prevent EM analysis attacks. The emitter generates an EM field which is constantly measured by the sensor. When an adversary brings a probe near this circuit, the change in EM-field will be detected by the EM sensor and the circuit will take the designed preventive measures to stop the EM analysis attack [50].

b. *High-frequency inductive voltage regulator*. EM analysis attacks work by measuring the EM field of the circuit. A prevention mechanism devised by Kar et al. [51] is that by emitting strong EM radiations, it can serve to blind the EM readings taken by the adversary. For this, a high-frequency inductive voltage regulator is used with a randomized control loop. This increased inductance acts as an EM emitter, making it difficult for an adversary to sample the accurate EM readings [51].

10.3.2.5 Acoustic Crypto-Analysis Attack

Acoustic Crypto-analysis attacks are a type of eavesdropping side-channel attacks. These attacks make use of sound information generated by IoT devices. By carefully analyzing the sounds of the IoT microprocessor or cryptographic circuit, adversaries can get critical information about processing data or keys.

a. *Minimizing sounds*. A simple mitigation technique to prevent acoustic analysis attacks is to minimize sounds an IoT device makes by using casings and coverings that absorb sounds [27]. Attackers will consequently be forced to use more expensive and sophisticated equipment to attack these devices [27]. Attackers run the risk of running out of resources used for an attack. Implementing this countermeasure is difficult for microprocessors as they need to have vents to remove excess heat [27]. Using heat sinks that are capable of dissipating heat without needing direct vents can help in this matter.

b. *Generating random parallel instructions*. One method for protecting against Acoustic Crypto-Analysis attack is to design the system in such a way that it randomly processes dummy parallel instructions [27]. In doing so, the system changes the sound and processing data correlation [27]. This will make it difficult for an adversary to correctly analyze the relation between acoustic data and processed data. This randomization process can also work in cryptographic calculations. By randomly processing inputs and then recovering the correct sequence while outputting data, randomized intermediate values can protect against acoustic analysis.

Figure 10.10 shows us how a parallel background load can shift the leakage frequency from the 35–38 kHz range to the range of 32–35 kHz [27].

c. *Generating parallel sound*. Another mitigation technique can be to analyze the sound emitted by an IoT device and to implement a mechanism in that IoT

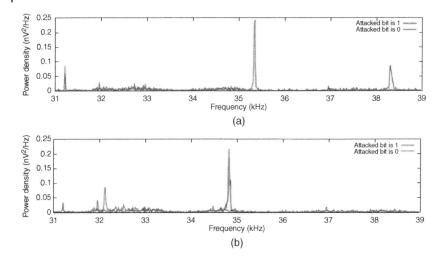

Figure 10.10 Acoustic measurement frequency spectra of the second modular exponentiation with and without constant load. (a) Without background load and (b) with background load [27].

device that generates sound within a similar spectrum. Creating parallel sound however is difficult and requires analyzing the sound produced by both inputs and operations. By generating parallel sounds of similar frequencies, adversaries will have a difficult time deciphering targeted noise from useful acoustic information from the hardware. To mask leakage through sounds generated by a device, a well-designed noise generator is required [27].

10.3.3 Integrated Circuits Security

ICs are the basic components of IoT devices. Secure provisioning and production of these ICs are a major security goal. Different security issues can arise if these chips are compromised. Some of the problems associated with IC are counterfeit IC, RE, and design theft. Preventive measures for most of these security issues are already discussed in the Prevention of Hardware Trojan section.

10.3.3.1 Countermeasures

During the production phase of IoT devices, many components are acquired from third-party vendors. Proper identification mechanisms to ensure the authenticity of these components are required to keep IoT devices secure. Proper data for each component must be maintained and scanned before use in the production process. If more than one supplier is involved, comparing the components with each other and testing can help distinguish the faulty components.

Physical Unclonable Functions (PUF) are hardware-based chips that differ from each other due to variations induced during the manufacturing process. These differences in microstructures are induced during the manufacturing process due to uncontrollable factors. Each PUF responds differently to the same input. This makes PUF ideal for authentication purposes [52]. These functions help resolve the problem of counterfeit IC and RE. One such technique that uses Memory-Based PUFs to secure IC against RE attacks is called *Hot Carrier Injection-Sense Amplifier* [52]. Even if two PUFs have identical structures, their responses will be different which makes cloning difficult even if the design of the circuit is leaked [52].

10.3.4 Radio Frequency Identification

RFID is a system used for identification. RFID systems consist of a tag and a reader and use an EM field to communicate between the two. Depending on the tag, the tag can respond to EM pulse from the reader from various distances. The reader on receiving the identification code can authenticate the tag. RFID has applications in almost every field of life. Due to the physical, power, and cost limitations of RFID tags, tags have simple circuits. These tags can be battery-powered. These characteristics of RFID systems make them susceptible to many attacks ranging from simple physical tampering attacks to more sophisticated side-channel attacks.

10.3.4.1 Physical Unclonable Function-based Authentication

PUF are hardware-based chips that differ from each other due to variations induced during the manufacturing process. These differences in microstructures are induced during the manufacturing process due to uncontrollable factors. This makes PUFs ideal for authentication purposes [52]. Each PUF has its unique response to the same stimulus/input, these responses depend on the variations in the microstructures of the circuit. This stimulus–response mechanism is called the challenge–response mechanism [53]. If an RFID tag is made using PUF it will be impossible for the adversary to clone this tag [54].

10.3.4.2 Preventing Physical Tampering Attacks (Enhancing Physical Security)

This kind of vandalism attacks can be prevented by proper security measures of the facilities and effective implementation of access control. By installing CCTV surveillance systems, alarm systems, and motion detectors, access to RFID readers can be secured [55]. Security of RFID readers is important as they hold the data regarding cryptographic algorithms and other authentication protocols. To increase security against attacks on RFID tags like tag removal attacks, these attacks can be mitigated by securing tags in the device in such a way that if someone tries to remove these tags, tags become useless or destroyed [18]. For example, using strong glue can make it impossible to remove the tag without damaging it. Moreover, having a mechanism that matches the tag with the device or access card to which the tag is attached and a centralized database of reported stolen tags can prevent tag tampering attacks [18].

10.3.4.3 Preventing Information Leakage

The information present on the tag or the reader is vulnerable to theft. If an adversary gets hold of a tag or a reader, they can easily gain access to the information stored within them. To prevent information leakage, minimal information should be stored on a tag or a reader [18]. The tag must only have a tag ID and other coded information stored that cannot disclose any information about the tag owner. The reader should have encrypted and coded information about tag data, and this information must be authenticated from a server or a third source which is at a more secure location [18]. Eavesdropping and information leakage can be minimized when minimal and encrypted data is transmitted during the authentication process [18].

10.3.4.4 Preventing Relay Attack

A relay attack uses an intermediate device to relay information of the original RFID tag to an unauthorized device. A relay is composed of two elements a proxy and a mole [56]. The mole is placed near the authorized tag and a proxy is placed near the reader. These two elements communicate together by a wired or a wireless link [56]. This unauthorized tag/proxy can be used by an adversary to enter a secure facility by relaying the information of a genuine tag to the emulated tag. Figure 10.11 shows how attackers can use a combination of a mole and a proxy to relay information of a payment card [56].

a. *Distance bounding protocol.* Distance bounding protocol authenticates the RFID tag by analyzing the time taken to send and receive the EM pulse by the RFID reader. The RFID reader is termed a verifier and the tag is termed a prover [56]. The verifier sends a series of encrypted messages to the prover, and the prover replies to these messages accordingly. The round trip time of each

Figure 10.11 Relay scenario in a queue. Source: Thevenon and Savry [56].

message is calculated by the verifier to assess the proximity of the prover. This helps to find if tag responding is a genuine tag and not an emulated one [56].

b. *Two-factor authentication*. Two-factor authentication uses a secondary parameter such as biometric verifiers to help prevent replay attacks. This solution is not feasible in automated systems and is complicated in the process of contactless identification.

Context-aware Selective Unlocking   Context-aware Selective Unlocking is a technique where the contextual information of the RFID tag is received by sensors, allowing the tag to only be unlocked when responding to the readers' challenges if it is in a specific state [57]. One such implementation of this technique is to monitor the posture of the RFID tag owner, based on this information, the card is unlocked [57]. This system uses a magnetometer to track tag orientation and an accelerometer to detect user motion. Data from these sensors are computed and stored in the tag to compare with a set of templates [57]. A decision based on the comparison is made to lock or unlock the card to respond to readers' queries.

One of the possible solutions for this problem is by adding a secondary identification parameter to the RFID tag. This can be a simple sound or noise emitting system, which can be turned off when the card is not in use. When an authorized user wants to authenticate their tag to the reader, they can turn on the random noise generating system and the reader can verify the tag and the noise to authenticate the user. Adding multiple identifying features to the tag holding cards or devices can help mitigate relay attacks.

Table 10.6 shows the costs associated with each countermeasure discussed in this chapter. This table can be used as a reference for designing an IoT device security mechanism. It will help in meeting the constraints of the specific IoT device.

10.4   Conclusion

The security of IoT devices is dependent on the security of each hardware component, software, and the cryptographic algorithm used in these devices. Attackers can target any of these components. As IoT devices become more ubiquitous within both personal use and private enterprise, it is certain that more sophisticated attacks will be seen in the future. As such, the task of securing an IoT device should start from the design of its first component until the end of its life cycle; from hardware design continuing into the deployment of these devices, security mechanisms have to be evaluated, implemented, and then reevaluated. Security of IoT devices is a continual process that will not stop as long as a device is in use. This holds true especially in the deployment of IoT devices for some critical applications. There are few important factors that should be kept in mind while

Table 10.6 Costs related to counter measure.

		Costs of counter measures			
Threats	**Counter measures**	**Costs**			
		Computational	**Financial**	**Battery**	**Time**
Hardware Trojans	Obscuring design information		*		
	Multi-mode operation		*		
	Removing extra space		*		
	Segregated production		*		*
Power analysis	Hiding (confusion)	*		*	*
	Dummy operations	*	*	*	
	Randomizing input data execution		*	*	
	Clock delays	*			*
	Masking	*		*	*
Timing attacks	Constant time operations				*
	Bucketing				*
	Random delays		*		*
	Pre-fetching T-tables in AES		*		
Electromagnetic analysis attacks	EM attack sensor		*		*
	High-frequency voltage regulator		*		*
Acoustic analysis attack	Minimizing sound		*		
	Generating parallel instructions	*		*	
	Generating parallel sound	*		*	
Abuse of tag	Physical unclonable function	*	*		
	Enhancing physical security		*		
	Preventing information leakage	*	*		*
Relays attacks	Distance bounding protocol	*	*		
	Two-factor authentication		*		*
	Context-aware selective unlocking	*	*	*	
Counterfiet ICs	Physical unclonable functions		*		
Reverse engineering					

designing and manufacturing an IoT device; as the devices can be compromised in the manufacturing process it is critical that companies ensure that their devices are free from HTs. Additionally, as the IoT is heavily reliant on communicating with the cloud, it is crucial that this communication is secure. Security best practices should continually be followed and evaluated as the threat to IoT devices is continually changing. The need for evaluation must be done in consideration of the unique limits of the hardware of IoT devices. While designing a security mechanism, one has to keep in mind the limitations of these devices. Mainly, processing power, battery, cost, and their often-remote deployments which make repair and maintenance difficult. While it is true that countermeasures have a cost related to them, whether it is a financial or computational cost, companies need to weigh these costs against their reputation, the balance of which dictates the overall security of an IoT device. To conclude, it is important to keep in mind that attacks are constantly evolving and new types of attacks altogether in the future will emerge. One must always remain vigilant when it comes to security and expect the unexpected to occur. With this in mind, countermeasures will need to evolve with time as well.

Acknowledgment

This chapter is the product of a team effort overseen by Dr. Saqib Hakak, each member of the team focused on a specific key area of research with Catherine Higgins and Lucas McDonald primarily focused on defining specific threats, while Muhammad Ijaz Ul Haq focused on addressing the countermeasures, after which the countermeasures were reviewed and edited by Catherine Higgins and Lucas McDonald.

References

1 Chen, K., Zhang, S., Li, Z. et al. (2018). Internet-of-things security and vulnerabilities: taxonomy, challenges, and practice. *Journal of Hardware and Systems Security* 2 (2): 97–110.
2 Sidhu, S., Mohd, B.J., and Hayajneh, T. (2019). Hardware security in IoT devices with emphasis on Hardware Trojans. *Journal of Sensor and Actuator Networks* 8 (3): 42.
3 Ji, W., Li, L., and Zhou, W. (2018). Design and implementation of a RFID reader/router in RFID-WSN hybrid system. *Future Internet* 10 (11): 106.
4 Landaluce, H., Arjona, L., Perallos, A. et al. (2020). A review of IoT sensing applications and challenges using RFID and wireless sensor networks. *Sensors* 20 (9): 2495.

5 Liu, L., Ma, Z., and Meng, W. (2019). Detection of multiple-mix-attack malicious nodes using perceptron-based trust in IoT networks. *Future Generation Computer Systems* 101: 865–879.

6 Mahmoud, R., Yousuf, T., Aloul, F., and Zualkernan, I. (2015). Internet of things (IoT) security: current status, challenges and prospective measures. *2015 10th International Conference for Internet Technology and Secured Transactions (ICITST)*, pp. 336–341. https://doi.org/10.1109/ICITST.2015.7412116.

7 Wang, P., Chaudhry, S., Li, L. et al. (2016). The internet of things: a security point of view. *Internet Research*. 26 (2) 337–359.

8 Butun, I., Österberg, P., and Song, H. (2019). Security of the internet of things: vulnerabilities, attacks, and countermeasures. *IEEE Communication Surveys and Tutorials* 22 (1): 616–644.

9 Sathyamoorthi, T., Vijayachakaravarthy, D., Divya, R., and Nandhini, M. (2014). A simple and effective scheme to find malicious node in wireless sensor network. *International Journal of Research In Engineering And Technology* 3 (02). 97–110.

10 Li, M., Koutsopoulos, I., and Poovendran, R. (2010). Optimal jamming attack strategies and network defense policies in wireless sensor networks. *IEEE Transactions on Mobile Computing* 9 (8): 1119–1133.

11 Zhang, G., Yan, C., Ji, X. et al. (2017). DolphinAttack: Inaudible voice commands. *Proceedings of the 2017 ACM SIGSAC Conference on Computer and Communications Security*, pp. 103–117.

12 Mao, J., Zhu, S., and Liu, J. (2020). An inaudible voice attack to context-based device authentication in smart IoT systems. *Journal of Systems Architecture* 104: 101696.

13 Naik, S. and Shekokar, N. (2015). Conservation of energy in wireless sensor network by preventing denial of sleep attack. *Procedia Computer Science* 45: 370–379.

14 Gallais, A., Hedli, T.-H., Loscri, V., and Mitton, N. (2019). Denial-of-sleep attacks against IoT networks. *2019 6th International Conference on Control, Decision and Information Technologies (CoDIT)*, pp. 1025–1030. IEEE.

15 Nie, X. and Zhong, X. (2013). Security in the internet of things based on RFID: issues and current countermeasures. *Proceedings of the 2nd International Conference on Computer Science and Electronics Engineering*, Atlantis Press.

16 Tsiropoulou, E.E., Baras, J.S., Papavassiliou, S., and Qu, G. (2016). On the mitigation of interference imposed by intruders in passive RFID networks. *International Conference on Decision and Game Theory for Security*, pp. 62–80. Springer.

17 Aghili, S.F., Mala, H., Kaliyar, P., and Conti, M. (2019). SecLAP: Secure and lightweight RFID authentication protocol for medical IoT. *Future Generation Computer Systems* 101: 621–634.

18 Mitrokotsa, A., Rieback, M.R., and Tanenbaum, A.S. (2010). Classification of RFID attacks. *Gen* 15693 (14443): 14.

19 Mitrokotsa, A., Rieback, M.R., and Tanenbaum, A.S. (2010). Classifying RFID attacks and defenses. *Information Systems Frontiers* 12 (5): 491–505.

20 Shamsoshoara, A., Korenda, A., Afghah, F., and Zeadally, S. (2019). A survey on hardware-based security mechanisms for internet of things. *arXiv preprint arXiv:1907.12525.*

21 Colombier, B. and Bossuet, L. (2014). Survey of hardware protection of design data for integrated circuits and intellectual properties. *IET Computers and Digital Techniques* 8 (6): 274–287.

22 Ender, M., Ghandali, S., Moradi, A., and Paar, C. (2017). The first thorough side-channel Hardware Trojan. In: *International Conference on the Theory and Application of Cryptology and Information Security*, (eds. Takagi Tsuyoshi, Peyrin Thomas), 755–780. Springer.

23 Rooney, C., Seeam, A., and Bellekens, X. (2018). Creation and detection of Hardware Trojans using non-invasive off-the-shelf technologies. *Electronics* 7 (7): 124.

24 Spreitzer, R., Moonsamy, V., Korak, T., and Mangard, S. (2017). Systematic classification of side-channel attacks: a case study for mobile devices. *IEEE Communication Surveys and Tutorials* 12. https://doi.org/10.1109/COMST. 2017.2779824.

25 Kocher, P., Jaffe, J., and Jun, B. (1999). Differential power analysis. *Annual International Cryptology Conference*, pp. 388–397. Springer.

26 Sayakkara, A., Le-Khac, N.-A., and Scanlon, M. (2019). Leveraging electromagnetic side-channel analysis for the investigation of IoT devices. *Digital Investigation* 29: S94–S103.

27 Genkin, D., Shamir, A., and Tromer, E. (2017). Acoustic cryptanalysis. *Journal of Cryptology* 30 (2): 392–443.

28 Kocher, P.C. (1996). Timing attacks on implementations of Diffie-Hellman, RSA, DSS, and other systems. *Annual International Cryptology Conference*, pp. 104–113. Springer.

29 Lo'ai, A.T. and Somani, T.F. (2016). More secure internet of things using robust encryption algorithms against side channel attacks. *2016 IEEE/ACS 13th International Conference of Computer Systems and Applications (AICCSA)*, pp. 1–6. IEEE.

30 Patranabis, S., Roy, D.B., Chakraborty, A. et al. (2019). Lightweight design-for-security strategies for combined countermeasures against side channel and fault analysis in IoT applications. *Journal of Hardware and Systems Security* 3 (2): 103–131.

31 Wei, L., Zhi, T., Dawu, G. et al. (2014). An effective differential fault analysis on the serpent cryptosystem in the internet of things. *China Communications* 11 (6): 129–139.

32 Aerabi, E., Papadimitriou, A., and Hely, D. (2019). On a side channel and fault attack concurrent countermeasure methodology for MCU-based byte-sliced cipher implementations. *2019 IEEE 25th International Symposium on On-Line Testing and Robust System Design (IOLTS)*, pp. 103–108. IEEE.

33 Kim, C.H. (2012). Differential fault analysis of AES: toward reducing number of faults. *Information Sciences* 199: 43–57.

34 Takarabt, S., Schaub, A., Facon, A. et al. (2019). Cache-timing attacks still threaten IoT devices. *International Conference on Codes, Cryptology, and Information Security*, pp. 13–30. Springer.

35 Venugopalan, V. and Patterson, C.D. (2018). Surveying the Hardware Trojan threat landscape for the internet-of-things. *Journal of Hardware and Systems Security* 2 (2): 131–141.

36 Cornell, N. and Nepal, K. (2017). Combinational Hardware Trojan detection using logic implications. *2017 IEEE 60th International Midwest Symposium on Circuits and Systems (MWSCAS)*, pp. 571–574. https://doi.org/10.1109/MWSCAS.2017.8052987.

37 Fujino, T., Kubota, T., and Shiozaki, M. (2017). Tamper-resistant cryptographic hardware. *IEICE Electronics Express* 14: 20162004. https://doi.org/10.1587/elex.14.20162004.

38 Standaert, F., Peeters, E., and Quisquater, J. (2005). On the masking countermeasure and higher-order power analysis attacks. *International Conference on Information Technology: Coding and Computing (ITCC'05) - Volume II*, Volume 1, pp. 562–567. https://doi.org/10.1109/ITCC.2005.213.

39 Popp, T., Mangard, S., and Oswald, E. (2007). Power analysis attacks and countermeasures. *IEEE Design Test of Computers* 24 (6): 535–543. https://doi.org/10.1109/MDT.2007.200.

40 Jerábek, S., Schmidt, J., Novotný, M., and Miškovský, V. (2018). Dummy rounds as a DPA countermeasure in hardware. *2018 21st Euromicro Conference on Digital System Design (DSD)*, pp. 523–528. https://doi.org/10.1109/DSD.2018.00092.

41 Boey, K.H., Lu, Y., O'Neill, M., and Woods, R. (2010). Random clock against differential power analysis. *2010 IEEE Asia Pacific Conference on Circuits and Systems*, pp. 756–759. https://doi.org/10.1109/APCCAS.2010.5774887.

42 Bayrak, A.G., Velickovic, N., Regazzoni, F. et al. (2013). An EDA-friendly protection scheme against side-channel attacks[C]. *Design, Automation & Test in Europe Conference & Exhibition (DATE)*.

43 Mangard, S., Popp, T., and Gammel, B.M. (2005). Side-channel leakage of masked CMOS gates. In: *Topics in Cryptology - CT-RSA* (ed. A. Menezes). Springer-Verlag Berlin Heidelberg. 351–365.

44 Mahanta, H.J., Barbhuiya, P., and Khan, A.K. (2016). A masking based RSA to resist power analysis attacks. *2016 International Conference on Computational Techniques in Information and Communication Technologies (ICCTICT)*, pp. 552–557. https://doi.org/10.1109/ICCTICT.2016.7514641.

45 Köpf, B. and Dürmuth, M. (2009). A provably secure and efficient countermeasure against timing attacks. *2009 22nd IEEE Computer Security Foundations Symposium*, pp. 324–335. https://doi.org/10.1109/CSF.2009.21.

46 Brumley, D. and Boneh, D. (2003). Remote timing attacks are practical. *Proceedings of the 12th Conference on USENIX Security Symposium - SSYM'03*, Volume 12, p. 1. USA. USENIX Association.

47 Jayasinghe, D., Ragel, R., and Elkaduwe, D. (2012). Constant time encryption as a countermeasure against remote cache timing attacks. *2012 IEEE 6th International Conference on Information and Automation for Sustainability*, pp. 129–134. https://doi.org/10.1109/ICIAFS.2012.6419893.

48 Alawatugoda, J., Jayasinghe, D., and Ragel, R. (2011). Countermeasures against Bernstein's remote cache timing attack. *2011 6th International Conference on Industrial and Information Systems*, pp. 43–48. https://doi.org/10.1109/ICIINFS.2011.6038038.

49 Bayat-Makou, P., Jahanian, A., and Reshadi, M. (2018). Security improvement of FPGA design against timing side channel attack using dynamic delay management. *2018 IEEE Canadian Conference on Electrical Computer Engineering (CCECE)*, pp. 1–4. https://doi.org/10.1109/CCECE.2018.8447738.

50 Miura, N., Fujimoto, D., Nagata, M. et al. (2015). EM attack sensor: concept, circuit, and design-automation methodology. *2015 52nd ACM/EDAC/IEEE Design Automation Conference (DAC)*, pp. 1–6. https://doi.org/10.1145/2744769.2747923.

51 Kar, M., Singh, A., Mathew, S. et al. (2018). Blindsight: blinding EM side-channel leakage using built-in fully integrated inductive voltage regulator.

52 Yamamoto, D., Takenaka, M., Sakiyama, K., and Torii N. (2014). A technique using PUFs for protecting circuit layout designs against reverse engineering. In: *Advances in Information and Computer Security. IWSEC* (ed. M. Yoshida and K. Mouri). Springer International Publishing. 158–173.

53 Braeken, A. (2018). Puf based authentication protocol for IoT. *Symmetry* 10: 352.

54 Devadas, S., Suh, E., Paral, S. et al. (2008). Design and implementation of PUF-based "unclonable" RFID ICs for anti-counterfeiting and security applications. *2008 IEEE International Conference on RFID*, pp. 58–64. https://doi.org/10.1109/RFID.2008.4519377.

55 Karygiannis, T., Eydt, B., Barber, G. et al. (2007). *Guidelines for Securing Radio Frequency Identi Cation (RFID) Systems: Recommendations of the National Institute of Standards and Technology.* NIST Special Publication 800-98, National Institute of Standards and Technology, Technology Administration U. S.

56 Thevenon, P.-H. and Savry, O. (2013). "Implementation of a countermeasure to relay attacks for contactless HF systems, " Radio frequency identification from system to applications, Mamun bin Ibne Reaz, IntechOpen. *DOI: 10,* p. 5772/53393.

57 Halevi, T., Li, H., Ma, D. et al. (2013). Context-aware defenses to RFID unauthorized reading and relay attacks. *IEEE Transactions on Emerging Topics in Computing* 1 (02): 307–318. https://doi.org/10.1109/TETC.2013.2290537.

Index

Security and Privacy in the Internet of Things: Architectures, Techniques, and Applications,
First Edition. Edited by Ali Ismail Awad and Jemal Abawajy.
© 2022 The Institute of Electrical and Electronics Engineers, Inc. Published 2022 by John Wiley & Sons, Inc.

Printed and bound by CPI Group (UK) Ltd, Croydon, CR0 4YY

16/04/2025

14658417-0002